Nadia Nedjah, Luiza de Macedo Mourelle, Mario Neto Borges,
Nival Nunes de Almeida (Eds.)

Intelligent Educational Machines

Studies in Computational Intelligence, Volume 44

Editor-in-chief
Prof. Janusz Kacprzyk
Systems Research Institute
Polish Academy of Sciences
ul. Newelska 6
01-447 Warsaw
Poland
E-mail: kacprzyk@ibspan.waw.pl

Further volumes of this series can be found on our homepage: springer.com

Vol. 26. Nadia Nedjah, Luiza de Macedo Mourelle (Eds.)
Swarm Intelligent Systems, 2006
ISBN 3-540-33868-3

Vol. 27. Vassilis G. Kaburlasos
Towards a Unified Modeling and Knowledge-Representation based on Lattice Theory, 2006
ISBN 3-540-34169-2

Vol. 28. Brahim Chaib-draa, Jörg P. Müller (Eds.)
Multiagent based Supply Chain Management, 2006
ISBN 3-540-33875-6

Vol. 29. Sai Sumathi, S.N. Sivanandam
Introduction to Data Mining and its Application, 2006
ISBN 3-540-34689-9

Vol. 30. Yukio Ohsawa, Shusaku Tsumoto (Eds.)
Chance Discoveries in Real World Decision Making, 2006
ISBN 3-540-34352-0

Vol. 31. Ajith Abraham, Crina Grosan, Vitorino Ramos (Eds.)
Stigmergic Optimization, 2006
ISBN 3-540-34689-9

Vol. 32. Akira Hirose
Complex-Valued Neural Networks, 2006
ISBN 3-540-33456-4

Vol. 33. Martin Pelikan, Kumara Sastry, Erick Cantú-Paz (Eds.)
Scalable Optimization via Probabilistic Modeling, 2006
ISBN 3-540-34953-7

Vol. 34. Ajith Abraham, Crina Grosan, Vitorino Ramos (Eds.)
Swarm Intelligence in Data Mining, 2006
ISBN 3-540-34955-3

Vol. 35. Ke Chen, Lipo Wang (Eds.)
Trends in Neural Computation, 2007
ISBN 3-540-36121-9

Vol. 36. Ildar Batyrshin, Janusz Kacprzyk, Leonid Sheremetor, Lotfi A. Zadeh (Eds.)
Preception-based Data Mining and Decision Making in Economics and Finance, 2006
ISBN 3-540-36244-4

Vol. 37. Jie Lu, Da Ruan, Guangquan Zhang (Eds.)
E-Service Intelligence, 2007
ISBN 3-540-37015-3

Vol. 38. Art Lew, Holger Mauch
Dynamic Programming, 2007
ISBN 3-540-37013-7

Vol. 39. Gregory Levitin (Ed.)
Computational Intelligence in Reliability Engineering, 2007
ISBN 3-540-37367-5

Vol. 40. Gregory Levitin (Ed.)
Computational Intelligence in Reliability Engineering, 2007
ISBN 3-540-37371-3

Vol. 41. Mukesh Khare, S.M. Shiva Nagendra (Eds.)
Artificial Neural Networks in Vehicular Pollution Modelling, 2007
ISBN 3-540-37417-5

Vol. 42. Bernd J. Krämer, Wolfgang A. Halang (Eds.)
Contributions to Ubiquitous Computing, 2007
ISBN 3-540-44909-4

Vol. 43. Fabrice Guillet, Howard J. Hamilton
Quality Measures in Data Mining, 2007
ISBN 3-540-44911-6

Vol. 44. Nadia Nedjah, Luiza de Macedo Mourelle, Mario Neto Borges, Nival Nunes de Almeida (Eds.)
Intelligent Educational Machines, 2007
ISBN 3-540-44920-5

Nadia Nedjah
Luiza de Macedo Mourelle
Mario Neto Borges
Nival Nunes de Almeida
(Eds.)

Intelligent Educational Machines

Methodologies and Experiences

With 52 Figures and 25 Tables

Springer

Nadia Nedjah
Universidade do Estado do Rio de Janeiro
Faculdade de Engenharia
Rua São Francisco Xavier
524, 20550-900 Maracanã
Rio de Janeiro, Brazil
E-mail: nadia@eng.uerj.br

Mario Neto Borges
Federal University of Sao Joao del Rei - UFSJ
Electrical Engineering Department (DEPEL)
Praca Frei Orlando 170 - Centro
CEP: 36.307-352
Sao Joao del Rei, MG, Brazil
E-mail: marionetoborges@uol.com.br

Luiza de Macedo Mourelle
Universidade do Estado do Rio de Janeiro
Faculdade de Engenharia
Rua São Francisco Xavier
524, 20550-900 Maracanã
Rio de Janeiro, Brazil
E-mail: ldmm@eng.uerj.br

Nival Nunes de Almeida
Universidade do Estado do Rio de Janeiro
Faculdade de Engenharia
Rua São Francisco Xavier
524, 20550-900 Maracanã
Rio de Janeiro, Brazil
E-mail: nival@eng.uerj.br

ISSN print edition: 1860-949X
ISSN electronic edition: 1860-9503
ISBN 978-3-642-07953-5
e-ISBN 978-3-540-44921-8

Springer is a part of Springer Science+Business Media
springer.com

Softcover reprint of the hardcover 1st edition 2007

Cover design: deblik, Berlin

To the memory of my father Ali and my beloved mother Fatiha,

Nadia Nedjah

To the memory of my beloved father Luiz and my mother Neuza,

Luiza de Macedo Mourelle

To all the lecturers using innovative pedagogical approaches for the development of the engineering courses

Mario Neto Borges

To my father and mother (in memory) for believing in education

Nival Nunes de Almeida

Preface

The Artificial Intelligence and Education is an ever-growing research area. It uses artificial intelligence tools and techniques to provide the foundations for intelligent tutoring and learning systems. The main research objective consists of using the computer-based technology to examine, discuss and fosters the processes of cognition, learning, teaching and so forth. It supports and develops innovative teaching methods and learning systems.

The goal of this volume has been to offer a wide spectrum of sample works developed in leading research throughout the world about innovative methodologies of artificial intelligence in education and foundations of engineering educational intelligent systems as well as application and interesting experiences in the field. The book should be useful both for beginners and experienced researchers interested in applying computational intelligence to the educational process.

In Chapter 1, which is entitled *A Framework for Building a Knowledge-Based System for Curriculum Design in Engineering*, the author demonstrates that the methodology of Knowledge-Based Systems can be applied to Curriculum Design. A general framework was devised to carry out the process of building this Knowledge-Based System with the domain of Curriculum Design divided into several subdomains. A strategy was developed and applied to investigate these interdependent subdomains independently.

In Chapter 2, which is entitled *A Web-based Authoring Tool for Intelligent Tutors: Blending Assessment and Instructional Assistance*, the authors cover the web-based architecture through which students and teachers interact and the Builder application, used internally to create the content. The authors report on the designing of the content and the evaluation of the assistance and assessment that the Assistment system provides.

In Chapter 3, which is entitled *Alife in the Classrooms: an Integrative Learning Approach*, the author applies Alife systems (Artificial Life) to the integrative learning of computation, biology, mathematics and scientific epistemology (methods, practices ...) in the classroom.

In Chapter 4, which is entitled *Pedagogic Strategies Based on the Student Cognitive Model Using the Constructivist Approach*, the authors assess whether it is possible to design pedagogic strategies based on models of conscience awareness and use them, by means of intelligent agents.

In Chapter 5, which is entitled *Tracing CSCL Processes*, the authors present the experience they have developed using a software tool called TeamQuest that includes activities that provide the opportunity for students to examine the performed task from different perspectives, needed to enable learners to make choices and reflect on their learning both individually and socially. The authors include a model that intend to evaluate the collaborative process in order to improve it based on the permanent evaluation and analysis of different alternatives.

In Chapter 6, which is entitled *Formal Aspects of Pedagogical Negotiation in AMPLIA System*, the authors present a pedagogical negotiation model developed for AMPLIA, an Intelligent Probabilistic Multi-agent Learning Environment. AMPLIA focuses on the formal aspects of the negotiation process, trying to abstract the most general characteristics of this process.

In Chapter 7, which is entitled *A New Approach to Meta-Evaluation Using Fuzzy Logic*, the authors present a methodology proposed and developed in Brazil for meta-evaluation that makes use of the concepts of fuzzy sets and fuzzy logic. It allows for the use of intermediate answers in the process of data collection. The proposed methodology allows: (i) the respondent to provide correct answers that indicate his (her) real understanding with regard to the response to a certain standard; (ii) to use linguistic rules provided by specialists, even with contradictory thinking; (iii) to deal with the intrinsic imprecision that exists in complex problems such as the meta-evaluation process.

In Chapter 8, which is entitled *Evaluation of Fuzzy Productivity of Graduate Courses*, the author intends to bring elements to the discussion of productivity measurement issues specially important in the evaluation of the Master courses. A main concern of CAPES, Brazil evaluation system is on not gauging results without taking into account the volume of resources applied. The fuzzy productivity measures are defined and key concepts of randomness and dependence are discussed. The author develop the procedures to be applied to completely quantify the productivity measures.

We are very much grateful to the authors of this volume and to the reviewers for their tremendous service by critically reviewing the chapters. The editors would also like to thank Prof. Janusz Kacprzyk, the editor-in-chief of the Studies in Computational Intelligence Book Series and Dr. Thomas Ditzinger from Springer-Verlag, Germany for their editorial assistance and excellent collaboration to produce this scientific work. We hope that the reader will share our excitement on this volume and will find it useful.

March 2006

Nadia Nedjah
Luiza M. Mourelle

State University of Rio de Janeiro
Brazil

Contents

List of Figures

List of Tables

1

A Framework for Building a Knowledge-Based System for Curriculum Design in Engineering

Mario Neto Borges

Department of Electrical Engineering, Federal University of Sao Joao Del Rei, Praca Frei Orlando, 170 - 36.307.904 Sao Joao Del Rei – MG –Brazil
marionetoborges@uol.com.br

Course designers in engineering may be subject matter specialists, but many lack knowledge of the principles and techniques for developing curricula. Also, as a broad generalisation, the literature on Curriculum Design tends to focus on theoretical issues and fails to address the practical needs of course designers in engineering. This work attempts to redress the balance in the direction of a more pragmatic approach to Curriculum Design by combining expertise about practical rules with theoretical knowledge within a Knowledge-Based System. This Knowledge-Based System thus represents an innovative approach to Curriculum Design.

Indeed the purpose of this Chapter is precisely to demonstrate that the methodology of Knowledge-Based Systems can be applied to Curriculum Design. A general framework was devised to carry out the process of building this Knowledge-Based System with the domain of Curriculum Design divided into several subdomains. A strategy was developed and applied to investigate these interdependent subdomains independently. This strategy is itself an important contribution to the theory of Knowledge Acquisition.

The study has shown that there is scope for acquiring expert knowledge and numerical rules about Curriculum Design and that these rules can be implemented in a portable, stand-alone system which can be used by experts and users. The tests carried out about the independent subdomains and about the system as a whole met the experts' and users' expectations.

1.1 Introduction

In the last decade or so there has been a move towards applying Knowledge-Based Systems to design and synthesis in different areas of knowledge, thus expanding its application boundaries [15, 6, 4]. These applications were previously almost confined to diagnosis and problem-solving which, by and large,

Mario Neto Borges: *A Framework for Building a Knowledge-Based System for Curriculum Design in Engineering*, Studies in Computational Intelligence (SCI) **44**, 1–22 (2007)
www.springerlink.com

dominated the development of Expert Systems from the 1960's to early 1980's [6]. Not only has this branch of Artificial Intelligence created a new generation of systems and been more realistically applied due to considerable advances in Computing and other related fields (such as Psychology, Mathematics), but, more importantly, it has found a novel application as a training tool and for expert advice [15], since it is a low cost and friendly way of disseminating knowledge and expertise.

In engineering, in general, and in engineering education, in particular, where computers are a natural tool of work for lecturers and students alike, a Knowledge-Based System seems to be able to play an important role in so far as it combines two essential ingredients, providing not only advice but also the knowledge and information that underpin the given advice. Moreover, the factual knowledge and information are readily available and easily accessible in a Knowledge-Based System application (which is not always the case in books and other related sources). This makes a Knowledge-Based System for consultancy about Curriculum Design particularly important for Course Designers who are the primary target of this study.

Higher Education, as a whole, and Curriculum Design, in particular, have been evolving relentlessly and more recently under vigorous economic pressures, namely budgetary constraints. The new world economic order and the escalating competitiveness among countries have called out for a multiskilled and better trained working force. Within this context the pressure on the Higher Education System has been intensified by the fact that some governments are promoting an increased undergraduate enrolment, as everywhere in the world the number of higher education applicants has been pushed upwards. Educationists have responded to these challenges by coming up with new proposals for education. In the United States of America, during the last decade, there have been some initiatives to approach this issue in Engineering Education such as Concurrent Engineering and Synthesis Coalition [16]. These initiatives primarily aim at striking a balance between efficiency and effectiveness in developing and running new courses in higher education. As a consequence, Course Designers have faced a huge task when designing new courses (or when updating the existent ones) to make these courses accessible to a larger number of students and more flexible in their implementation while, at the same time, not reducing the quality of the learning process. This task of developing new courses has been particularly difficult due to a lack of practical guidelines on curriculum design, though theoretical publications abound in this area. This whole context may have brought about the increasing status of Curriculum Design and Engineering Education departments within Universities nowadays.

From the point of view of many countries, the lack of expertise in Curriculum Design and the difficulty in accessing practical advice, added to dwindling resources for education, exacerbate the problem. In Brazil, for instance, this has been manifested through the issue of Student-Staff Ratios (SSRs). On the one hand, there are the private universities which claim to have kept SSRs relatively high but have found little space for research even on how to improve their own curricula. On the other hand, there are the Federal universities (and some State ones as in the case of São Paulo state) where the SSRs are still low - yet under pressure - but which have been at the forefront of the research achievements in the country. This is an international problem [10, 13].

Many of the Course Designers lack the necessary background in curriculum theory, being often industrial specialists recruited to education or new doctors on their subject area. Course Designers, world-wide, are urged by their governments or their institutions to approach the design of curricula in a systematic way from the outset. While trying to fulfil the needs for a new course, should they not take into consideration the strengths and weaknesses of their own institution before starting to make decisions about the curriculum? When a new curriculum is being designed the decision-making process is bound to address specific areas of Curriculum Design such as: (a) the aims and outcomes; (b) the structure of the course; (c) the identification of the curriculum content; (d) the teaching and learning strategy and (e) the assessment methods. They are therefore discussed briefly in the paragraphs below. Would it not also be important to consider that, although all these areas related to the curriculum may be looked at independently, they should be treated as part of an integrated domain, as a Systematic Approach to Curriculum Design would suggest?

With that in mind, in the present study, an Introduction to Curriculum Design (embodied in a Knowledge-Based System) aims to prime the Course Designers on those relevant issues which may lead to a successful development of the curriculum (which - otherwise - would lack coherence and consistency). This introduction is followed by a close look at the identification of the Curriculum Content. This is one of the areas in Curriculum Design that needs to be approached taking into account an institution's resources and capabilities. That is to say, among the several alternative ways of setting about identifying the content (which should be incorporated in the curriculum), how does one decide the alternative that best matches the institution's staff profile and other resources with the educational requirements which sparked off the need for a new course? Also in this regard, how could Course Developers profit from being able to determine the appropriate Method for identifying the curriculum content for their institutions through the use of a Knowledge-Based System and, at the same time, learn what expertise should be developed amongst the staff as far as the Methods of developing curricula are concerned?

Course Structure is also a major concern for Course Designers. The National Curriculum Guidelines for Engineering being implemented in Brazil at the moment is a case in point. The structure of a course is subject to all sorts of different pressures. On the one hand, the costs of staff time to teach in a particular course structure and resources for laboratories are examples of factors which impose fewer staff-student contact hours and less practical (hands-on) learning experiences. On the other hand: (a) the continuous expansion of the content to be covered (with a soaring number of new topics and techniques brought into the curriculum); (b) the flexibility of the curriculum and options to be made available to students and (c) also a more student-centred approach being recommended in higher education; are features that requires more staff time and more physical resources to run courses in engineering. Course Designers are, consequently, the ones responsible for designing a structure which takes account of both sets of pressures. Would therefore a Knowledge-Based System, incorporating guidelines and knowledge in this area of Curriculum Design, prove helpful to Course Designers by addressing, among others, the above mentioned issues?

There has recently been a new trend in Higher Education towards focusing on the students' achievements rather than on the learning process itself; the roots of this trend are in the Learning Outcomes theory [11]. It seems to be accepted that this new approach to Curriculum Design suits best the present requirements of the market place, given that it maps out what graduates in engineering are expected to be able to do after having undertaken their learning experience. This theory suggests that a degree course could be described in terms of its learning outcomes. It assumes that achievement is defined by the successful demonstration of learning outcomes and that a group of Learning Outcomes Statements defines the coherent learning experience characterised as a course unit. Would a Knowledge-Based System embodying knowledge and expertise in this area prove useful to Course Designers?

Moreover, it is self-evident that the strategy for assessing students' learning cannot be neglected throughout the curriculum design process and plays a crucial role in the Learning Outcomes theory. It has not always been clear how the assessment procedures actually measure the broad range of qualities expected from engineering graduates. There have been strong criticisms of assessment procedures on the grounds that they lack a coherent theoretical framework and are arbitrary [8]. As a result, Course Designers are put under considerable pressure to come up with a Scheme of Assessment which, within the limits of an institution's resources, represents an appropriate and acceptable measure of achievement and from which students can benefit throughout the learning process. Therefore, could it be that practical rules and knowledge, which give advice in this context, whilst taking into consideration particular institutional needs (which may differ from institution to institution), prove to be an essential tool for Course Designers?

It can be seen from these sub-areas of Curriculum Design that there is a synergy among the issues discussed which cannot be overlooked if the design of a new curriculum is to succeed in being coherent, efficient and effective. In this Chapter an innovative Knowledge-Based System is presented which embodies not only very practical rules to handle intelligent curriculum principles and concepts, but also the knowledge and information underlying these rules in all these areas of Curriculum Design mentioned above. Thus, this novelty in curriculum design for engineering degree courses represents an alternative access to curriculum theory, particularly in the areas mentioned, for those who develop the curriculum (the End-users). The assumption is that this

Knowledge-Based System can make a great impact on education by enabling course design to be improved and by resulting in the better education and training of students for their future professional careers.

The diagram in figure 1.1 shows a simple flow chart for a Curriculum Design process in a typical Institution of Higher Education. The proposition underlying the present work is that the grey area in the diagram represents the points within the process where a Knowledge-Based System can come in to assist the End-user in the task of developing course proposals and producing course documents. It is in this phase of the Curriculum Design process that this Knowledge-Based System can have great impact.

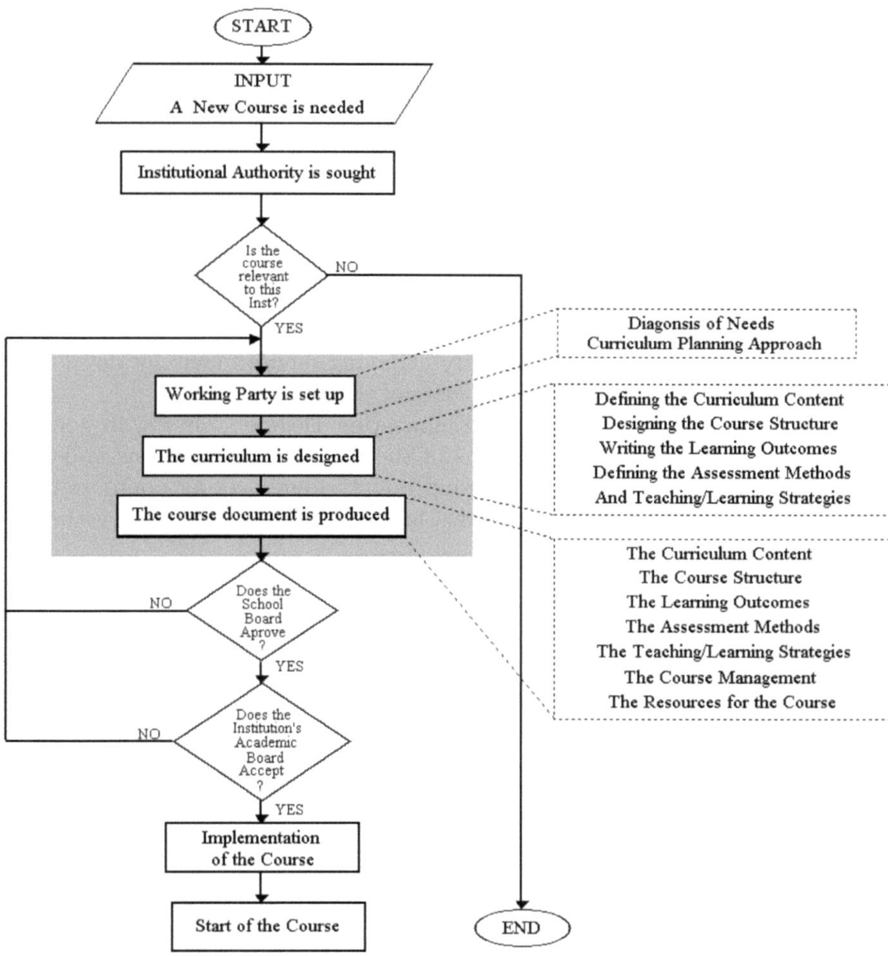

Fig. 1.1. A Simplified Flow Chart for the Curriculum Design Process

1.2 Rational

The rationale and impetus for this study came from the following observations and findings:

- The design and development of engineering degree courses at most Universities and Institutions of Higher Education worldwide have been carried out by Course Leaders and Course Committees (comprising lecturers and students) who, very often, do not have training and expertise in principles of Curriculum Design. Their expertise is based only on their previous educational experience [1].
- There has been a lack of financial resources for higher education and there are also conflicting views over the use of these resources. On the one hand, the academic community is claiming that the financial support is rather scarce and does not meet the needs of a realistic curriculum for Higher Education. On the other hand, the educational funding agencies (governments) are saying that the funds made available should be used more efficiently and even suggest that the Higher Education System should accommodate more students (that is, have a higher Student-Staff Ratio) .
- Furthermore, it is emphasised here that there have been rapid advances in science and technology which must be taken into account in the engineering curriculum. Therefore, the curriculum should have a flexible and dynamic structure to be able to, at least, try to keep up with these fast changes. The more important point is that the delivery of the curriculum should prepare graduates to cope with the rapidly changing environment by developing and enhancing transferable and personal skills.
- As a result of these factors the Engineering Degree Courses, in general, have not fulfilled the expectations of the academic institutions and have not satisfied the needs of employers and the engineering community at large [5]. In other words, they have failed to address adequately the national needs.

1.3 Aims

In order to cope properly with the problems mentioned above, the intention for this Chapter is to pursue the following aims:

1. To demonstrate that the methodology of Knowledge-Based Systems can be applied to Curriculum Design.
2. To present a framework for developing a Knowledge-Based System in Engineering which can provide Course Designers with both:
3. a set of intelligent curriculum principles, which can be quickly accessed and,
4. specific advice in their particular contexts, which takes account of local needs and suits their specific requirements.

1.4 The Framework

This chapter recognises and justifies the need for several experts in the process of building a Knowledge-Based System in the context of *Curriculum Design*. The alternatives of building these systems for a complex domain are discussed in depth and a framework, which addresses this issue, is presented in detail. The methodology is to divide the domain of *Curriculum Design* into separate subdomains and to have a Domain-expert and several Subdomain-experts working independently with a Knowledge Engineer. The framework presented in this chapter minimised the problem of conflict of expertise by restrictions on the subdomain *boundaries* and *limits* through the concepts of *input* and *output* variables. The concepts of *boundaries* and *limits* are explained as being the bases of the framework and they have been devised to keep the integration and size of the knowledge base under control. The *input* and *output* variables for the domain of *Curriculum Design* are presented in full. They are the major driving force behind the knowledge elicitation sessions carried out in the subdomains investigated throughout the development of this Knowledge-Based System. It is also shown that this novel approach has addressed successfully the issues of verification and validation of Knowledge-Based Systems by presenting an iterative and interactive knowledge acquisition process in which End-users, the Subdomain-experts and Domain-expert play a very important role in contributing to build the final system.

1.5 Alternative Approaches to Knowledge Engineering

It is well recognised that Knowledge-Based Systems are often built in an "ad hoc" way with a limited theoretical base [9, 4] and that the knowledge elicitation is clearly identified as being the "bottleneck" of the process. Underlying this problem is the fact that experts do not usually have their knowledge and expertise structured in the way required by the knowledge acquisition process and very often different techniques are needed to elicit such knowledge and expertise in terms of practical rules. This has proved to be very much the case in the domain of *Curriculum Design*. Moreover, the verification and validation of Knowledge-Based Systems are currently major concerns regarding Knowledge-Based System technology. This chapter presents a framework devised to address these issues over a particular application. In this application of Knowledge-Based System methodology to *Curriculum Design* it is most unlikely that a single expert is able to cover the domain; consequently the reconciliation of expertise is a problem that needs to be addressed.

In simple Knowledge-Based System applications it is possible to find a single expert who has both the knowledge and the available time to provide all the expertise required to build a Knowledge-Based System. In such cases this expert usually has the competence and authority to carry out, together with a knowledge engineer, the verification and validation of the system, thus enabling the expert to judge the system's development and behaviour. On the other hand, in other more complex and extensive applications no single expert is likely to be able to cover appropriately all aspects of the area (domain) or commit enough time to the project development or both. These are, consequently, the cases in which more than one expert should be involved in the development of the system, thus requiring a very well-planned strategy to effectively approach the problem. The strategy must be clearly devised well in advance in order to both divide the domain into sub-areas, which could be dealt with independently, and to identify who would be the expert to provide the necessary expertise in each sub-area. At this point, a rough estimate of the amount of time, that will be required to develop each subdomain and consequently the time required from each expert, should be worked out.

Curriculum Design is seen as a poorly structured task which may not have an optimum solution and its specification is rather difficult. This is understandable in an area which does not always produce exact results and good evidence can be provided for the use of different theoretical approaches. Furthermore, some principles in this area would have different interpretations when used in different contexts. What is required, therefore, is a clear picture of the context to be combined with practical educational principles and rules, if an effective curriculum is to be designed for a particular context. This application focuses on educational principles about *Curriculum Design* for engineering degree courses rather than on curriculum content. Consequently, educationists play a major role in its development. Nevertheless, the area is so vast that one particular expert may feel more comfortable discussing, say, Student Assessment than Course Structure or vice-versa. In other words, within the area of *Curriculum Design* expertise in the sub-areas may be better provided by specific experts who have acquired experience in one particular sub-area throughout their professional practice. This is, therefore, an archetypal case for a Knowledge-Based System application using several experts. This is mainly because advantage can be taken of the Knowledge-Based System feature which allows expertise from different experts to be combined with one another in the same knowledge base.

If the application requires several experts there is still a possibility that the knowledge engineer could work with only one expert. This expert would provide all the information to the knowledge engineer, not only from the expert's own knowledge but also by obtaining the necessary expertise through discussions with other experts. Scott et al. in [12] correctly point out that the knowledge acquisition is more streamlined in this approach than with multiple experts and the organisation of the project is usually facilitated for the knowledge engineer. However, there are still two hurdles associated with this approach which are often difficult to overcome. Firstly, this expert has to commit a huge amount of time to working with the knowledge engineer and to acquiring the complementary expertise from colleagues. Such time is hardly ever available in the case of experts. Secondly, the knowledge acquired from other experts may be biased or inadequate for the purposes of Knowledge-Based Systems. In other words, experts are not knowledge engineers.

It is not uncommon to see applications requiring multiple experts in which the task to be embodied in the Knowledge-Based System is performed in teams of several experts, each of whom, usually specialises in a particular phase or part of the task. This automatically suggests that it is advisable to divide the domain into sub-areas to be dealt with separately rather than to attempt to work with the whole team in a group discussion. As far as knowledge acquisition is concerned, group work has been discouraged as the literature suggests [2, 12] on the grounds that it is time-consuming and ineffective with severe problems of communication.

The alternative strategy, which worked well in the domain of *Curriculum Design* in the present study, is that the Knowledge Engineer worked with the several experts independently (one at a time), even if the different sub-areas within the domain to be investigated were overlapping somewhat. There was one expert who delineated the whole domain and was named here as the Domain-expert; those who provided expertise in each sub-area of knowledge were named Subdomain-experts. This strategy is discussed in full in the following section.

Having decided to embark on this approach, a great deal of time had to be spent by the Knowledge Engineer building productive working relationships with a number of Subdomain-experts. Also, some constraints were issued by the Domain-expert to keep the growth of the knowledge base under control. This is where the concepts of *boundaries* and *limits* in this framework played a pivotal role in the methodology. Moreover, a variety of knowledge elicitation techniques were necessary to suit different experts placing great demands on the Knowledge Engineer. In addition, conflict of expertise was likely to take place since experts seldom agree among themselves. These issues were addressed as shown in the following section, where a framework is presented which focuses on:

- the strategy adopted to delimit the domain as far as the human expertise is concerned;
- the knowledge acquisition in *Curriculum Design*;
- the procedures for verification and validation of the subdomains implemented.

1.6 Methodology

1.6.1 Domain Delineation

The strategy adopted to approach these issues can be visualised in figure 1.2. Concerning the full specification of the domain, the Domain-expert delineated the whole domain in a knowledge engineering exercise which is discussed in full

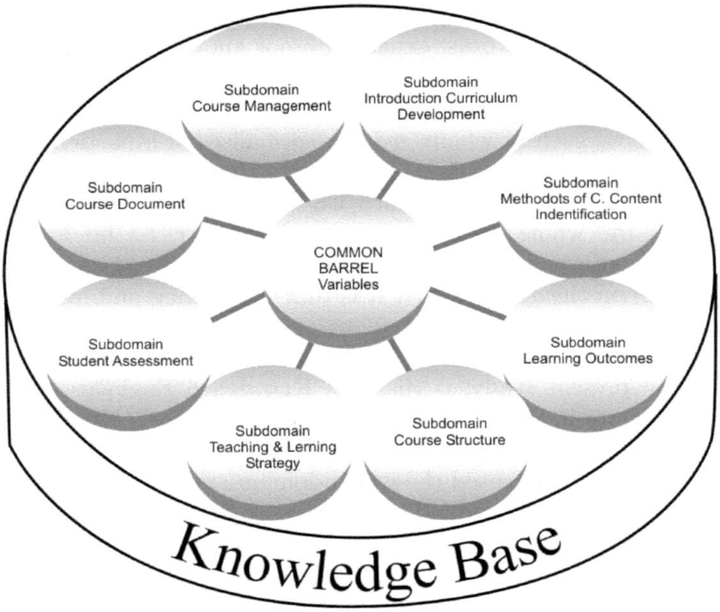

Fig. 1.2. Strategy for domain delineation. Knowledge Base for Curriculum Design

in section "The Specification of the Domain". The domain, represented in the diagram by the external barrel, was divided into subdomains as extensively suggested in the literature [17, 2]. These subdomains within the broader area (small barrels) were, therefore, decided by the Domain-expert.

The Domain-expert also defined the *boundaries* and *limits* for each subdomain. *Boundaries* (see Figure 1.3(a)) means how much the subdomains were allowed to expand against each other resulting in a pre-defined overlapping area and *limits* (see Figure 1.3(b)) means how much the subdomains could inflate themselves within the domain. Making the point in another way, the former is related to the interfacing of the subdomains, whereas the latter is related to their individual size and consequently to the size of the eventual knowledge base. The *boundaries* and other variables previously defined acted as *inputs* for each subdomain. On the basis of these inputs the Subdomain-experts were then able to decide how the outcomes of their particular area of expertise could be defined (*outputs*). Subdomain-experts had to produce all of the *outputs* plus any extra ones which they considered relevant. They could also use any *input* from the central bubble named *Common Barrel* (Figure 1.2) and add extra facts although they were not allowed to create other *input* variables. All variables inside the *Common Barrel* were generated as *outputs* from one subdomain or from the Domain-expert.

The Knowledge Engineer played a very important part in this strategy, not only by acquiring knowledge from different sources but also by linking the

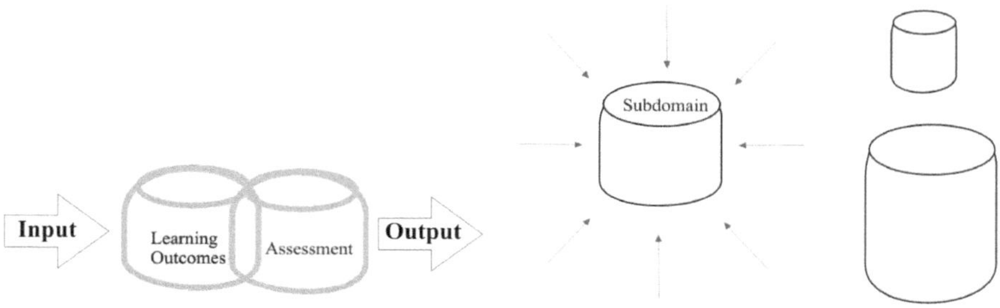

(a) (b)

Fig. 1.3. (a) - Subdomain Boundaries, (b) - Subdomain Limits

Domain-expert to the Subdomain-experts making the variables from other subdomains and those available in the *Common Barrel* accessible to the latter. This is emphasised in figure 1.2 by the connection between the *Common Barrel* and each subdomain. The Knowledge Engineer also brought to the Domain-expert's arbitration the conflict of expertise from different Subdomain-experts in the few instances where they happened. The Domain-expert decided how to settle the argument so that the Knowledge Engineer could implement the consensus expertise. The important aspect was that all Subdomain-experts were warned in advance and prepared to defer to Domain-expert decisions before conflicts had arisen.

1.6.2 Subdomain Investigation

Regarding the knowledge elicitation from the Subdomain-experts: each of them, individually, was given the initial guidelines of the project in a first session and was asked to outline (as they wished) the subdomain. Then, in the following sessions, an iterative process was adopted in which the Knowledge Engineer began by acquiring top level knowledge and then proceeded in a cyclical fashion to probe further and deeper into the details of the expert's skills within the subdomain. As a result of these sessions the knowledge could be represented in a conceptual model for the subdomain and checked with the Subdomain-expert. Once the Subdomain-expert had agreed with the conceptual model, the subdomain was then implemented in a prototype and appraised by the Subdomain-expert in cyclical sessions until all modifications and refinements lived up to the Subdomain-expert's expectations. This was carried out as simulation cases and demonstrations, both as a form of a further knowledge elicitation session to extend the existing knowledge base by broadening the scope of the system and as a form of verification refining the knowledge base. These simulation sessions for review identified inaccuracies or omissions, thus allowing for the knowledge acquisition plan to be refined in order to reflect the appropriate expertise, given that the Subdomain-expert was acting on simulated data. This cyclical procedure is shown on figure 1.4.

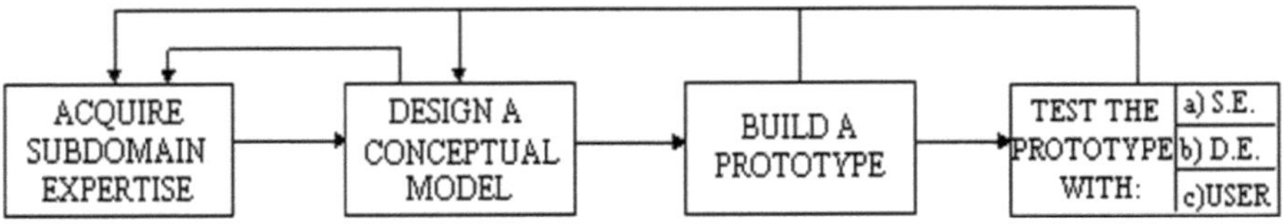

Fig. 1.4. Iterative approach for knowledge acquisition with the Subdomain-experts

1.6.3 Verification, Validation and Acceptance

Throughout this work the concepts of verification and validation were interpreted in the manner defined in [7]. To sum up, verification was related to the question "Are we doing the project right?" and validation concerns "Are we doing the right project?". Regarding verification, which was the Subdomain-experts' responsibility, the technique used was that which fostered expert-computer interaction throughout the elicitation procedure, thereby making sure that the Subdomain-experts were continuously assessing the system being built (see Figure 1.4). It also allowed the Domain-expert to oversee the prototypes in order to keep the size of the whole knowledge base under control.

Validation, from the experts' point of view, enjoyed a privileged position in this methodology in so far that it was seen as a cross-reference device between Domain-expert and Subdomain-experts. This methodology strengthened some components of the validation of the system described in [7] such as:

- competency (the quality of the system's decisions and advice compared with those obtained from sources of knowledge other than the Subdomain-experts);
- completeness (the system could deal with all the pre-defined *inputs* and *outputs* for its domain) and
- consistency (the knowledge base must produce similar solutions to similar problems, the knowledge must not be contradictory).

It must be said that, due to their experience and background, the experts (the Domain and Subdomain-experts) found it hard to see the problems and difficulties faced by the End-user when running a consultation. It is also important to point out that as far as End-users were concerned, they would be able to comment on the acceptability and facilities offered by the system, not on the expertise embodied in the knowledge base. The fact that prototypes for each subdomain were quickly built made it possible to test them for acceptability and usability involving the End-user in simulation sessions. These sessions were designed to identify the modifications in the performance of the prototypes that would provide incentives for using the Knowledge-Based System as a tool for assistance in *Curriculum Design*. Sometimes it was necessary to extend the task of the system or to include a new function in order for the prototype to meet the End-user's expectations. As a result, the End-user requirements and impressions were incorporated in early stages of the system's development, which was highly conducive to the improvement of the user interface for the eventual Knowledge-Based System. This methodology was applied to ensure, as far as possible, that the Knowledge-Based System would eventually be used and, more importantly, useful for the targeted Course Designers. In short, this framework presents a synergy among the participants, who have developed and would use the Knowledge-Based System, that strongly contributed to its successful completion and utility.

1.7 Knowledge Acquisition in Curriculum Development

1.7.1 The Specification of the Domain

An appropriate delineation of the domain is crucial in this methodology and this is a difficult task for a non-trivial Knowledge-Based System such as *Curriculum Design*. The domain delineation started with some Orientation Interviews [2]. In this initial inquiry stage the Knowledge Engineer gathers the general knowledge needed to prepare for the development of the system. The subsequent detailed investigation stage, which is required to obtain the specific knowledge that enables the system to perform its tasks, starts with the Card-Sort Technique [3].

The Card-Sort Technique was regarded as the most convenient knowledge elicitation technique to suit the stage of domain delineation and proved to be very effective, in that it was both naturally easy for the Domain-expert and helpful for the Knowledge Engineer to become acquainted with the domain knowledge. The method also favoured dividing the functionality of the Knowledge-Based System.

The Card-Sort Technique consisted of first presenting to the Domain-expert a set of individual cards on which concepts related to the domain of *Curriculum Design* were written down. The cards were spread out at random and the Domain-expert was told to group together the concepts into as many small groups as possible. The Domain-expert could add or exclude cards (concepts) at any time so as to better represent the knowledge in that domain. By thinking aloud, the Domain-expert's reasoning was recorded as the rationale behind the grouping was verbalised. The Domain Expert was then interviewed and asked to label the different groups and to accommodate them in larger groups. The structure and temporal relationships among these large groups were acquired at the next stage. The example for *Curriculum Design* demonstrated that in the first round of card-sorting (a video-recorded session) forty-eight cards (one for each domain concept) were presented to the Domain-expert. Subsequently, several audio recorded structured interviews took place to tailor the concepts to the point of view of the Domain-expert. This resulted in 114 concepts which described the domain of *Curriculum Design* (from the 48 initially presented by the Knowledge Engineer) and they are presented in Table 1.1 and Table 1.2.

After grouping the concepts, the groups were presented to the Domain-expert in a video recorded interview. By using the Teaching Back technique [6] the validity of the domain structure was checked. The Teaching Back technique was used partially to undertake an investigation of the concepts and of the domain structure, and to narrow the focus of the analysis. At this stage, the Knowledge Engineer taught the concepts and structure of the domain back to the Domain-expert, who was the final judge, in the Domain-expert's terms and to the Domain-expert's satisfaction. When it was agreed that the Knowledge Engineer was following the procedure in the Domain-expert's way, then it could be said that both shared the same concepts. Having agreed that the same procedure had been followed, the Knowledge Engineer asked the Domain-expert to give an explanation of how the domain structure was constructed. The teachback procedure continued until the Domain-expert was satisfied with the Knowledge Engineer's version. It could be said that at this point the Knowledge Engineer had understood the Domain-expert. Thus, to summarise, firstly all concepts were shared and then understanding was achieved.

1.7.2 Definition of the Inputs and Outputs for the Subdomains

At this point, the concepts were divided into eight subdomains: *Introduction to Curriculum Design, Methods for Curriculum Content Identification, Learning Outcomes, Course Structure, Teaching and Learning Strategies, Student Assessment, Course Documentation and Course Management.* Each is represented by a barrel in Figure 1.2. Together they make up the knowledge base for this application. Having defined the subdomains above, the knowledge engineering process was focused on deciding what variables would comprise the *inputs* and *outputs* for each subdomain and for the *Common Barrel.* These *inputs* and *outputs* are described in Tables 1.3–1.11. The investigation carried out with the Domain-expert for this phase of the methodology required 12 hours of knowledge elicitation sessions and a variety of the knowledge elicitation techniques mentioned above.

1.7.3 Software Architecture

Once the domain had been delineated, the subdomains had been identified and the *inputs* and *outputs* had been defined, the Knowledge Engineer, who was becoming an expert, could begin the investigation of the subdomains. However, some decisions had first to be made on the software construction concerning the use of: a) a prototype technique and b) the computer system. Despite the fact that several tools and techniques for building Knowledge-Based Systems are becoming available in the last decade [4], the decisions were made as follows:

1. **The Prototype Technique** - As far as the software architecture was concerned, among the strategies available for building a Knowledge-Based

Table 1.1. Domain Delineation

Systematic Planning	Planning Approach		**Curriculum Planning**
Expedient Planning			
Piecemeal Planning			
Introspection	Identification Methods	Diagnosis of Need	**Rationale**
Curriculum Enquiry			
DACUM Method			
DELPHI Technique			
Task Analysis			
Resources	Institutional Response		
Background			
Development			
Admission	Legislation		
Regulation			
Vocationalism	Balance	Policy	**Philosophical and Ideological Bases**
Liberalism			
Course Aims	Course Guidelines		
Course Goals			
Modular Course	Course Content Structure		**Curriculum Design**
Unit Course			
Grouped Course			
Integrated Curriculum			
Single Subject-based			
Time-Table Structure	Course Organization		
Course Duration			
Full Time			
Part Time			
Sandwich			
Personal Qualities	Educational Taxonomy	Competencies	**Learning Outcomes**
Knowledge			
Skills			
Student Performance	Standards		
Conditions of Learning			
Performance Criteria			
Process Centred	Pedagogical Approach		**Teaching/Learning Strategies**
Product Centred			
Student Centred			
Teacher Centred			
Lecturing	Learning Activities	Teaching and Learning Methods	
Tutorial			
Practical			
Unsupervised Work			
Teaching Techniques			
Study Skills			
Educational Technology			
Teac. Resource/Materials			

Table 1.2. Domain Delineation (Cont.)

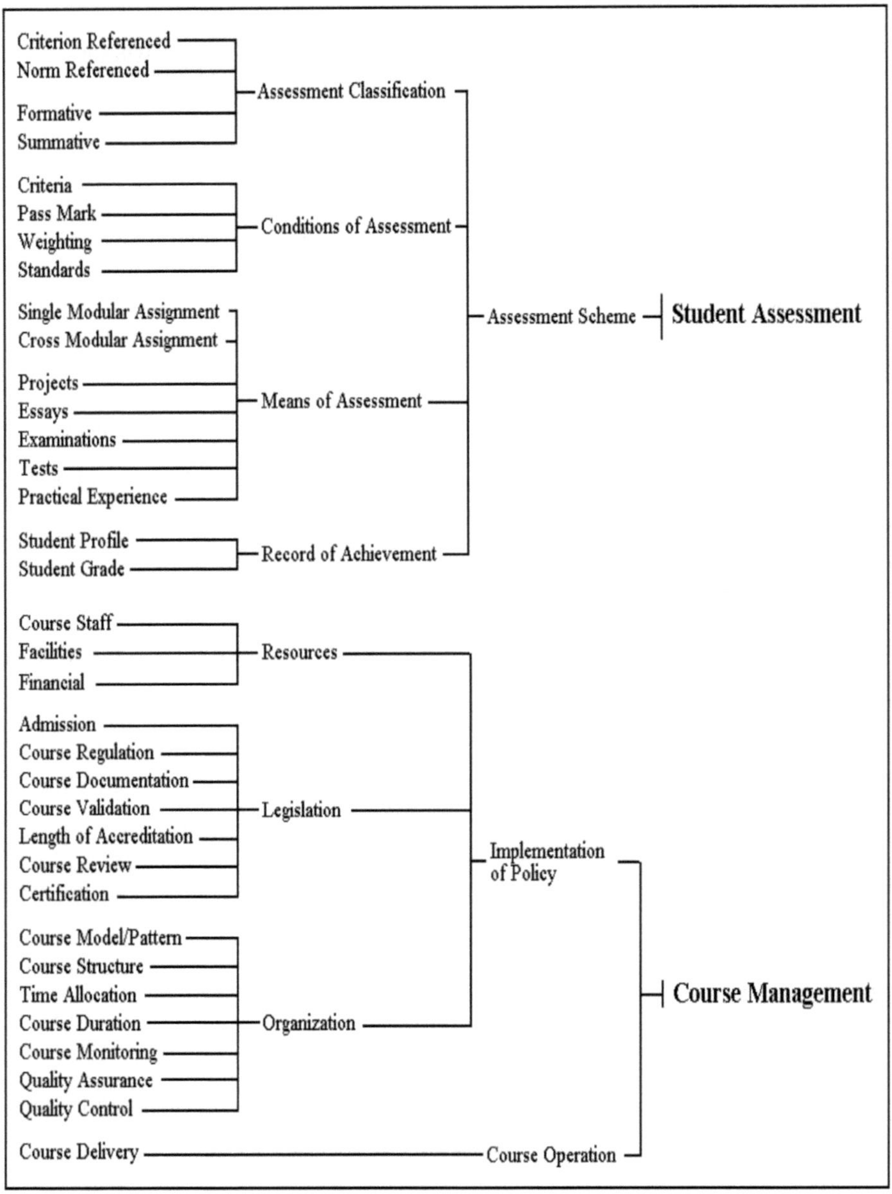

Table 1.3. Subdomain delineation (*Inputs* and *Outputs*)

INTRODUCTION TO CURRICULUM DEVELOPMENT		
I N P U T S		O U T P U T S
(from) the Subdomains	(from) the Common Barrel	
Course Content	Course Rationale	Major Concepts and Facts
Course Structure	Curric. Planning approach	SWOT Analysis
	Laws and Regulations	Development Time
	Resources	Personnel Structure
		Staff Team - Working Party

Table 1.4. Subdomain delineation (*Inputs* and *Outputs*) (Cont.)

METHODS OF CURRICULUM CONTENT IDENTIFICATION		
I N P U T S		O U T P U T S
(from) the Subdomains	(from) the Common Barrel	
Course Structure	Course Rationale	Major Concepts and Facts
Organisation of Units	Curric. Planning approach	Staff Profile
Duration of Units	Laws and Regulations	Methods Curric. Identific.
	Course Level and Focus	Curriculum Budget

Table 1.5. Subdomain delineation (*Inputs* and *Outputs*) (Cont.)

C O U R S E S T R U C T U R E		
I N P U T S		O U T P U T S
(from) the Subdomains	(from) the Common Barrel	
Course Content	Course Rationale	Major Concepts and Facts
Learning Outcomes	Curric. Planning approach	Course Design and Pattern
Scheme of Assessment	Laws and Regulations	Time Allocation (activities)
	Course Level and Focus	Progression System
	Award(s)	Attendance Pattern
		Entry Requirements

Table 1.6. Subdomain delineation (*Inputs* and *Outputs*) (Cont.)

L E A R N I N G O U T C O M E S		
I N P U T S		O U T P U T S
(from) the Subdomains	(from) the Common Barrel	
Course Content	Course Rationale	Major Concepts and Facts
Organisation of Units	Curric. Planning approach	A Method to Design LOSs
Duration of Units	Laws and Regulations	Educational Taxonomy
Scheme of Assessment	Course Level and Focus	How to Write LOSs
		How to Use LOSs

Table 1.7. Subdomain delineation (*Inputs* and *Outputs*) (Cont.)

S T U D E N T A S S E S S M E N T		
I N P U T S		O U T P U T S
(from) the Subdomains	(from) the Common Barrel	
Objectives and Outcomes	Course Rationale	Major Concepts and Facts
Educational Taxonomy	Course Level and Focus	A Scheme of Assessment
Number of Units	Laws and Regulations	Conditions of Assessment
Duration of Units	Award System	Means of Assessment
Number Progress Points		Record of Achievement

Table 1.8. Subdomain delineation (*Inputs* and *Outputs*) (Cont.)

TEACHING AND LEARNING STRATEGIES		
I N P U T S		O U T P U T S
(from) the Subdomains	(from) the Common Barrel	
Course Content	Course Rationale	Major Concepts and Facts
Course Structure	Curric. Planning approach	Pedagogical Approaches
Learning Outcomes	Course Level and Focus	Learning Activities
Scheme of Assessment		Teaching Methods

Table 1.9. Subdomain delineation (*Inputs* and *Outputs*) (Cont.)

C O U R S E D O C U M E N T		
I N P U T S		O U T P U T S
(from) the Subdomains	(from) the Common Barrel	
Course Content	Course Rationale	Major Concepts and Facts
Course Structure	Laws and Regulations	Units Specification
Learning Outcomes	Course Level and Focus	Time Allocation
Scheme of Assessment	Award(s)	Course Duration
Teach. Learning Strategies		Course Pattern
		Certification

Table 1.10. Subdomain delineation (*Inputs* and *Outputs*) (Cont.)

C O U R S E M A N A G E M E N T		
I N P U T S		O U T P U T S
(from) the Subdomains	(from) the Common Barrel	
Course Content	Course Rationale	Major Concepts and Facts
Course Structure	Laws and Regulations	Course Organisation
Learning Outcomes	Course Level and Focus	Course Monitoring
Scheme of Assessment	Award System	Quality Assurance/Control
Teach. Learning Strategies		Course Delivery
		Course Review

Table 1.11. Common Variables (Cont.)

C O M M O N B A R R E L		
VARIABLES AVAILABLE		
Award System	Curric. Planning approach	Length of Accreditation
Course Level and Focus	Facilities Required	Philosophical Basis
Course Rationale	Implementation Policy	Resources
Course Validation	Laws and Regulations	

System, the Incremental Prototype was chosen on the grounds that it is a feasible and convenient alternative for developing this system [14]. Prototyping is described in [1]. Regarding this study, the Incremental Prototype, was intended not only to provide a basis for establishing that a Knowledge-Based System would be a viable alternative approach for assisting *Curriculum Design*, but also to demonstrate to what extent the implementation of this project would be achievable in terms of Knowledge-Based System applications (see Figure 1.4). Having built a prototype for each subdomain, each prototype constituted a quite separate knowledge base which needed to be linked to the others in the software implementation. The methodology described in the previous section and the Incremental Prototype technique enabled the linking and amalgamating of these prototypes (in the software program) to be carried out in a rigorous manner. The linked prototypes were embodied in an Expert System Shell which runs on a PC environment, thus making the Knowledge-Based System more friendly and accessible to the targeted End-user.

2. **The Computer System** - The computer system built took advantage of a shell to facilitate the package demonstration during the knowledge acquisition sessions. A rule-based shell was chosen in which the control uses Forward Chaining to reach the goals. This means that it works from known facts supplied by the End-user or available in the knowledge base, towards desired goals through a scanning procedure. This is particularly suited to an advisory system as it is the Knowledge-Based System developed in this study. Furthermore, the use of an Expert System shell is recommended in the literature [15, 4]. The shell facilities include a DOS Extender for very large applications (that is, more than 1000 rules). Also, it has External Interfaces for other different software packages.

Despite the facilities incorporated in the shell, a great amount of work had to be carried out by the Knowledge Engineer in so far as the representation of the knowledge and rules elicited from the Domain and Subdomain-experts relied on the Knowledge Engineer. In addition to the facilities of a shell, a software program had to be written in a shell compatible language in order to represent the knowledge acquired in each subdomain and to integrate them in the same knowledge base. The fact that an Expert System Shell is particularly suitable for use in a PC environment made the

Knowledge-Based System developed in this study compatible with every PC (IBM compatible) machine.

1.8 Conclusions

The methodology of Domain-expert and Subdomain-experts in the present study has worked well in terms of being acceptable and has overcome the issue of expert conflict. The Subdomain-experts were happy with the methodology used and mentioned that a prior definition of *inputs* and *outputs* for their subdomains had been helpful particularly because this information told them where to start and where to finish. This method has placed a considerable burden on the Knowledge Engineer and this, in turn, justified not using the Domain-expert as a Knowledge Engineer with the Subdomain-experts. The concept of *boundaries* and *limits* has been successful in this area of *Curriculum Design* where the knowledge was not immediately available in rule form. The knowledge engineering for the different experts has used diverse methods, but the use of a single Knowledge Engineer and the Incremental Prototype technique have proved successful.

The contribution to knowledge that comes from this study can be seen in:

1. Formulating a novel approach to *Curriculum Design* by:
   a) drawing on a knowledge base formed not only by the theory (from the available literature) but also by the expertise (from human experts), which has never before been recorded in a formal, logical and numerate way;
   b) defining (and implementing in the knowledge base) the subdomains in *Curriculum Design* that are suitable for applying the Knowledge-Based System approach with the challenging task of synthesising knowledge in those areas;
   c) testing the Knowledge-Based System techniques for encapsulating the related knowledge and expertise in the domain of *Curriculum Design*.
2. Expanding the frontiers of Knowledge-Based Systems by developing a methodology of Knowledge Acquisition (Domain-Subdomain and Limits-Boundaries);
3. Making available an Intelligent Software Package which contains a comprehensive knowledge base in curriculum design and which can be used as a sophisticated form of information retrieval in the area of *Curriculum Design*;
4. Developing a tool for training in curriculum principles and concepts using a tutorial element which is a software structure built in parallel with the software for the main consultation of the Knowledge-Based System.

A relevant feature of this Knowledge-Based System is that it is user-friendly. This means that the End-users do not need to know how the system was built or to know any computer language in order to run a consultation and interact with the system. The Knowledge-Based System, as usual, contains the necessary general principles (or generic knowledge) and the End-user provides the specific data so that the expertise is formed by the interaction of the two bases (that is, the System's knowledge base and the End-user's data base).

The literature survey carried out for this study has revealed that no attempt has been made so far to provide rules or advice which take into consideration the particular context of institutions of higher education when discussing *Curriculum Design* issues. This is exactly the point where this Knowledge-Based System comes in to help with the design and development of curricula. The system can state the intelligent practical rules which organise and represent the knowledge in the domain of *Curriculum Design* in a way that is entirely novel. Not only does the Knowledge-Based System embody the basic alternative methods and procedures in *Curriculum Design* but, more importantly, it has stored into the knowledge base the expertise to answer adequately specific questions related to *Curriculum Design.*

Another consideration is that most of the work in the area of *Curriculum Design* has been concerned with theoretical issues. The novelty of the present study is that it incorporates a close investigation of practical issues and uses a rule-based system in a systematically planned approach to *Curriculum Design.* The writing of intelligent practical rules, which organise and represent the factual and expert knowledge in this field, bridges the existing knowledge gap between theory and practice. This novelty is reflected in the fact that the Subdomain-experts were at first satisfied with an esoteric, unspecific level of goal, but with further Knowledge Elicitation they produced the explicit, specific information as expected in the overall methodology. This framework can be applied to other domains which have the same level of complexity.

References

1. Borges, M. N., The Design and Implementation of a Knowledge-Based System for Curriculum Development in Engineering, Ph.D. Thesis, The University of Huddersfield, UK, 1994.
2. Firlej, M. and Hellens, D., Knowledge Elicitation, a practical handbook, London: Prentice Hall, 1991.
3. Gammack, J. G., Different Techniques and Different Aspects on Declarative Knowledge, In Kidd, A. L. (Ed.) Knowledge Acquisition for Expert Systems – A practical handkook, New York: Plenun Press, 1987.
4. Gennari, J. H. et al., The evolution of Protégé: An environment for Konwledge-Based Systems Development, International Journal of Human-Computer Studies, pp. 1–32, 2003.
5. Giorgetti, M. F., Engineering and Engineering Technology Education in Brazil, European Journal of Engineering Education 18(4), pp. 351–357, 1993.
6. Jackson, P., Introduction to expert systems, Wokinghan: Addison-Wesley, 1990.

7. Lydiard, T. J., Overview of current practice and research initiatives for the verification and validation of KBS, The Knowledge Engineering Review 7(2), pp. 101–113, 1992.
8. Otter, S., Learning Outcomes in Higher Education, A Development Project Report, UDACE, Employment Department, 1992.
9. Plant, R.T., Rigorous approach to the development of knowledge-based systems, Knowledge-Based Systems 4(4), pp. 186–196, 1991.
10. Psacharopoulos, G., Higher education in developing countries: the scenario of the future, Higher Education 21(1), pp. 3–9, 1991.
11. Robertson, D., Learning Outcomes and Credits Project, UDACE Project, The Liverpool Polytechnic, 1991.
12. Scott, A. C. and Clayton, J. E. and Gibson, E. L., A Practical Guide to Knowledge Acquisition, New York: Addison- Wesley, 1991.
13. Shute, J.C.M and Bor, W.V.D., Higher education in the Third World: status symbol or instrument for development, Higher Education 22(1), pp. 1–15, 1991.
14. Teo, A. S. and Chan, M. and Chunyan Miao, Incorporated framework for incremental prototyping with object-orientation, Proceedings of IEEE International Conference on Engineering Management Conference, Vol. 2, pp 770–774, 2004.
15. Vadera, S., Expert System Applications, Wilmslow: Sigma Press, 1989.
16. Watson, G. F., Refreshing curricula, IEEE Spectrum, pp. 31–35, March 1992.
17. Wiig, K., Expert Systems - A manager's guide, Geneva: International Labour Office, 1990.

2

A Web-based Authoring Tool for Intelligent Tutors: Blending Assessment and Instructional Assistance

Leena Razzaq[1], Mingyu Feng[1], Neil T. Heffernan[1], Kenneth R. Koedinger[2], Brian Junker[2], Goss Nuzzo-Jones[1], Michael A. Macasek[1], Kai P. Rasmussen[1], Terrence E. Turner[1], and Jason A. Walonoski[1]

[1] Worcester Polytechnic Institute, 100 Institute Road, Worcester, Massachusetts, USA
[2] Carnegie Mellon University, 5000 Forbes Ave., Pittsburgh, Pennsylvania, USA
assistments@wpi.edu

Middle school mathematics teachers are often forced to choose between assisting students' development and assessing students' abilities because of limited classroom time available. To help teachers make better use of their time, a web-based system, called the Assistment system, was created to integrate assistance and assessment by offering instruction to students while providing a more detailed evaluation of their abilities to the teacher than is possible under current approaches. An initial version of the Assistment system was created and used in May, 2004 with approximately 200 students and over 1000 students currently use it once every two weeks. The hypothesis is that Assistments can assist students while also assessing them. This chapter describes the Assistment system and some preliminary results.

2.1 Introduction

Limited classroom time available in middle school mathematics classes compels teachers to choose between time spent assisting students' development and time spent assessing students' abilities. To help resolve this dilemma, assistance and assessment are integrated in a web-based system called the Assistment[3] System that will offer instruction to students while providing a more detailed evaluation of their abilities to the teacher than is possible under current approaches. The plan is for students to work on the Assistment website for about 20 minutes per week. As building intelligent tutoring systems can be

[3] The term Assistment was coined by Kenneth Koedinger and blends Assisting and Assessment.

Leena Razzaq et al.: *A Web-based Authoring Tool for Intelligent Tutors: Blending Assessment and Instructional Assistance*, Studies in Computational Intelligence (SCI) **44**, 23–49 (2007)
www.springerlink.com

very costly [15], the Office of Naval Research provided funding to reduce those costs. We reported on the substantial reductions in time needed to build intelligent tutoring systems with the tools we have built.[4] The Assistment system is an artificial intelligence program and each week when students work on the website, the system "learns" more about the students' abilities and thus, it can hypothetically provide increasingly accurate predictions of how they will do on a standardized mathematics test. The Assistment System is being built to identify the difficulties individual students - and the class as a whole - are having. It is intended that teachers will be able to use this detailed feedback to tailor their instruction to focus on the particular difficulties identified by the system. Unlike other assessment systems, the Assistment technology also provides students with intelligent tutoring assistance while the assessment information is being collected.

An initial version of the Assistment system was created and tested in May, 2004. That version of the system included 40 Assistment items. There are now over 700 Assistment items. The key feature of Assistments is that they provide instructional assistance in the process of assessing students. The hypothesis is that Assistments can do a better job of assessing student knowledge limitations than practice tests or other on-line testing approaches by using a "dynamic assessment" approach. In particular, Assistments use the amount and nature of the assistance that students receive as a way to judge the extent of student knowledge limitations.

The rest of this chapter covers 1) the web-based architecture we used that students and teachers interact with, 2) the Builder application that we use internally to create this content and finally 3) a report on the designing of the content and the evaluation of the assistance and assessment that the Assistment system provides.

2.2 The Extensible Tutor Architecture

The eXtensible Tutor Architecture (XTA) is a framework that controls the interface and behaviors of our intelligent tutoring system via a collection of modular units. These units conceptually consist of a curriculum unit, a problem unit, a strategy unit, and a logging unit. Each conceptual unit has an abstract and extensible implementation allowing for evolving tutor types and content delivery methods. The XTA is represented by the diagram given in Figure 1, illustrating the actual composition of the units. This diagram shows

[4] This research was made possible by the US Dept of Education, Institute of Education Science, "Effective Mathematics Education Research" program Grant No. R305K03140, the Office of Naval Research Grant No. N00014-03-1-0221, NSF CAREER award to Neil Heffernan, and the Spencer Foundation. Author Razzaq was funded by the National Science Foundation under Grant No. 0231773. All the opinions in this article are those of the authors, and not those of any of the funders.

the relationships between the different units and their hierarchy. Within each unit, the XTA has been designed to be highly flexible in anticipation of future tutoring methods and interface layers. This was accomplished through encapsulation, abstraction, and clearly defined responsibilities for each component. These software engineering practices allowed us to present a clear developmental path for future components. That being said, the current implementation has full functionality in a variety of useful contexts.

2.2.1 Curriculum Unit

The curriculum unit can be conceptually subdivided into two main pieces: the curriculum itself, and sections. The curriculum is composed of one or more sections, with each section containing problems or other sections. This recursive structure allows for a rich hierarchy of different types of sections and problems.

Progress within a particular curriculum, and the sections of which is it composed, are stored in a progress file - an XML meta-data store that indexes into the curriculum and the current problem (one progress file per student per curriculum).

The section component is an abstraction for a particular listing of problems. This abstraction has been extended to implement our current section types, and allows for future expansion of the curriculum unit. Currently existing section types include "Linear" (problems or sub-sections are presented in linear order), "Random" (problems or sub-sections are presented in a pseudo-random order), and "Experiment" (a single problem or sub-section is selected pseudo-randomly from a list, the others are ignored). Plans for future section types include a "Directed" section, where problem selection is directed by the student's knowledge model [2].

2.2.2 Problem Unit

The problem unit represents a problem to be tutored, including questions, answers, and relevant knowledge-components required to solve the problem. For instance pseudo-tutors are a hierarchy of questions connected by correct and incorrect answers, along with hint messages and other feedback. Each of these questions are represented by a problem composed of two main pieces: an interface and a behavior.

The interface definition is interpreted by the runtime and displayed for viewing and interaction to the user. This display follows a two-step process, allowing for easy customization of platform and interface specification. The interface definition consists of high-level interface elements ("widgets"), which can have complex behavior (multimedia, spell-checking text fields, algebra parsing text fields). These "high-level" widgets have a representation in the runtime composed of "low-level" widgets. "Low-level" widgets are widgets common to many possible platforms of interface, and include text labels, text

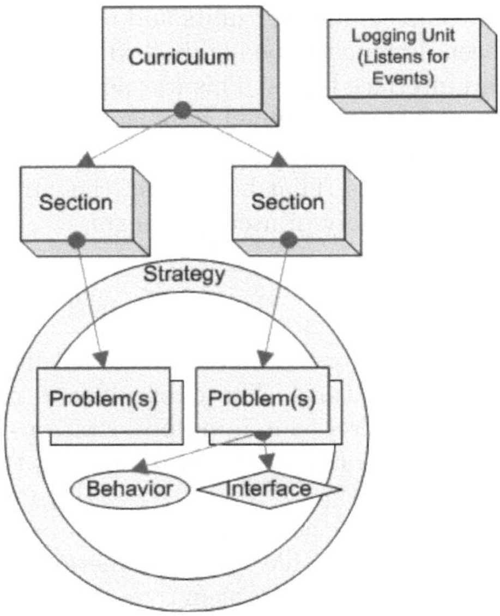

Fig. 2.1. Abstract unit diagram

fields, images, radio buttons, etc. These "low-level" widgets are then consumed by an interface display application. Such applications consume "low-level" widget XML, and produce an interface on a specific platform. The event model (described below) and relationship of "high-level" to "low-level" widgets allow a significant degree of interface customizability even with the limitations of HTML. Other technologies, such as JavaScript and streaming video are presently being used to supplement our interface standard. Future interface display applications are under consideration, such as Unreal Tournament for Warrior Tutoring [12], and Macromedia Flash for rich content definition.

The behaviors for each problem define the results of actions on the interface. An action might consist of pushing a button or selecting a radio button. Examples of behavior definitions are state graphs, cognitive model tracing, or constraint tutoring, defining the interaction that a specific interface definition possesses. To date, state graph or pseudotutor definitions have been implemented in a simple XML schema, allowing for a rapid development of pseudo tutors [16]. We have also implemented an interface to the JESS (Java Expert System Shell) production system, allowing for full cognitive model behaviors. A sample of the type of cognitive models we would wish to support is outlined in Jarvis et al [9]. The abstraction of behaviors allows for easy extension of both their functionality and by association their underlying XML definition.

Upon user interaction, a two-tiered event model (see Figure 2) is used to respond to that interaction. These tiers correspond to the two levels of widgets described above, and thus there are "high-level" actions and "low-level" actions. When the user creates an event in the interface, it is encoded as a "low-level" action and passed to the "high-level" interface widget. The "high-level" interface widget may (or may not) decide that the "low-level" action is valid, and encode it as a "high-level" action. An example of this is comparing an algebra text field (scripted with algebraic equality rules) with a normal text field by initiating two "low-level" actions such as entering "3+3" and "6" in each one. The algebra text field would consider these to be the same "high-level" action, whereas a generic text field would consider them to be different "high-level" actions. "High-level" actions are processed by the interpreted behavior and the interface is updated depending on the behavior's response to that action. The advantage of "high-level" actions is that they allow an interface widget or content developer to think in actions relevant to the widget, and avoid dealing with a large number of trivial events.

2.2.3 Strategy Unit

The strategy unit allows for high-level control over problems and provides flow control between problems. The strategy unit consists of tutor strategies and the agenda. Different tutor strategies can make a single problem behave in different fashions. For instance, a scaffolding tutor strategy arranges a number of problems in a tree structure, or scaffold. When the student answers the root problem incorrectly, a sequence of other problems associated with that incorrect answer is queued for presentation to the student. These scaffolding problems can continue to branch as the roots of their own tree. It is important to note that each problem is itself a self-contained behavior, and may be an entire state graph/pseudo-tutor, or a full cognitive tutor.

Other types of tutor strategies already developed include message strategies, explain strategies, and forced scaffolding strategies. The message strategy displays a sequence of messages, such as hints or other feedback or instruction. The explain strategy displays an explanation of the problem, rather than the problem itself. This type of tutoring strategy would be used when it was already assumed that the student knew how to solve the problem. The forced scaffolding strategy forces the student into a particular scaffolding branch, displaying but skipping over the root problem. The concept of a tutor strategy is implemented in an abstract fashion, to allow for easy extension of the implementation in the future. Such future tutor strategies could include dynamic behavior based on knowledge tracing of the student log data. This would allow for continually evolving content selection, without a predetermined sequence of problems.

This dynamic content selection is enabled by the agenda. The agenda is a collection of problems arranged in a tree, which have been completed or have been queued up for presentation. The contents of the agenda are operated upon by the various tutor strategies, selecting new problems from sections

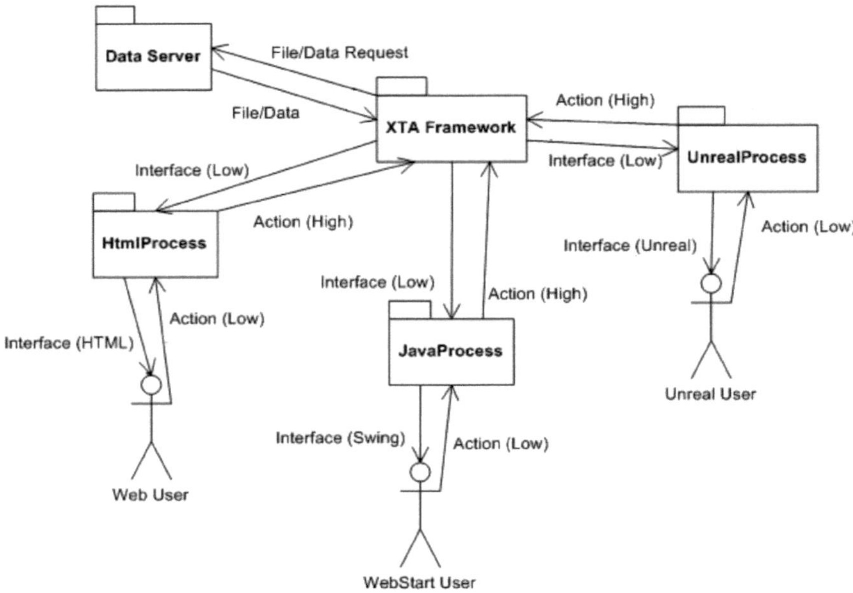

Fig. 2.2. Network architecture and event model diagram.

(possibly within sections) within a curriculum to append and choosing the next problem to travel to [7].

2.2.4 Logging Unit

The final conceptual unit of the XTA is the logging unit with full-featured relational database connectivity. The benefits of logging in the domain of ITS have been acknowledged, significantly easing data mining efforts, analysis, and reporting [14]. Additionally, judicious logging can record the data required to replay or rerun a user's session.

The logging unit receives detailed information from all of the other units relating to user actions and component interactions. These messages include notification of events such as starting a new curriculum, starting a new problem, a student answering a question, evaluation of the student's answer, and many other user-level and framework-level events.

Capturing these events has given us an assortment of data to analyze for a variety of needs. User action data captured allows us to examine usage-patterns, including detection of system gaming (superficially going through tutoring content without actually trying to learn) [7]. This data also enables us to quickly build reports for teachers on their students, as well as giving a complete trace of student work. This trace allows us to replay a user's session, which could be useful for quickly spotting fundamental misunderstandings on the part of the user, as well as debugging the content and the system itself (by attempting to duplicate errors).

The logging unit components are appropriately networked to leverage the benefits of distributing our framework over a network and across machines. The obvious advantage this provides is scalability.

2.2.5 System Architecture

The XTA provides a number of levels of scalability. To allow for performance scalability, care was taken to ensure a low memory footprint. It is anticipated, based on simple unit testing, that thousands of copies of the XTA could run on a single machine. More importantly, the individual units described above are separated by network connections (see Figure 2). This allows individual portions of the XTA to be deployed on different computers. Thus, in a server context, additional capacity can be added without software modification, and scalability is assured.

The runtime can also transform with little modification into a client application or a server application instantiated over a web server or other network software launch, such as Java WebStart. Both types of applications allow for pluggable client interfaces due to a simple interface and event model, as described in the interface unit. A client side application contains all the network components described above (Figure 2) as well as content files required for tutoring, and has the capacity to contact a remote logging unit to record student actions. Running the XTA in a server situation results in a thin client for the user (at present either HTML or Java WebStart), which operates with the interface and event model of the server. Thus the server will run an instance of the XTA for every concurrent user, illustrating the need for a small memory footprint. The XTA instances on the server contact a centralized logging unit and thus allow for generated reports available through a similar server [4].

2.2.6 Methods

The XTA has been deployed as the foundation of the Assistments Project [12]. This project provides mathematics tutors to Massachusetts students over the web and provides useful reports to teachers based on student performance and learning. The system has been in use for three years, and has had thousands of users. These users have resulted in over 1.3 million actions for analysis and student reports [4]. To date, we have had a live concurrency of approximately 50 users from Massachusetts schools. However, during load testing, the system was able to serve over 500 simulated clients from a single J2EE/database server combination. The primary server used in this test was a Pentium 4 with 1 gigabyte of RAM running Gentoo Linux. Our objective is to support 100,000 students across the state of Massachusetts. 100,000 students divided across 5 school days would be 20,000 users a day. Massachusetts schools generally have 7 class periods, which would be roughly equivalent to supporting 3,000 users concurrently. This calculation is clearly based on estimations, and it should be noted that we have not load tested to this degree.

Tutors that have been deployed include scaffolding state diagram pseudo-tutors with a variety of strategies (see Figure 3 for a pseudo-tutor in progress). We have also deployed a small number of JESS cognitive tutors for specialized applications. It should be noted that the tutors used in the scaling test described above were all pseudo-tutors, and it is estimated that a much smaller number of JESS tutors could be supported.

In summary, the launch of the XTA has been successful. The configuration being used in the Assistments project is a central server as described above, where each student uses a thin HTML client and data is logged centrally. The software has been considered stable for several months, and has been enthusiastically reviewed by public school staff. Since September 2004, the software has been in use at least three days a week over the web by a number of schools across central Massachusetts. This deployment is encouraging, as it demonstrates the stability and initial scalability of the XTA, and provides significant room to grow.

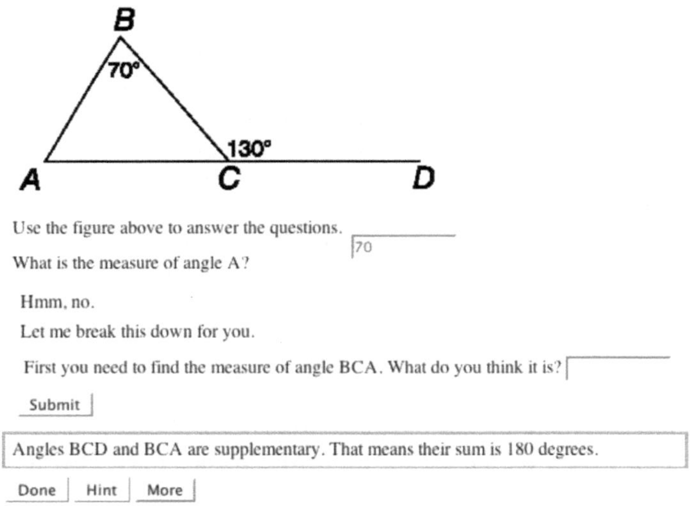

Fig. 2.3. Pseudo-tutor in progress.

The larger objective of this research was to build a framework that could support 100,000 students using ITS software across the state of Massachusetts. We're encouraged by our initial results from the Assistments Project, which indicate that the XTA has graduated from conceptual framework into a usable platform (available at http://www.assistments.org). However, this test of the software was primarily limited to pseudo-tutors, though model-tracing tutors are supported. One of the significant drawbacks of model-tracing tutors in a server context is the large amount of resources they consume. This resource consumption would prohibit scaling to the degree that is described in our results. A partial solution to this might be the support of constraint-based tutors [10], which could conceivably take fewer resources, and we are presently exploring this concept. These constraint tutors could take the form of a simple JESS model (not requiring an expensive model trace), or another type of scripting language embedded in the state-graph pseudo-tutors.

Other planned improvements to the system include dynamic curriculum sections, which will select the next problem based on the student's performance (calculated from logged information). Similarly, new tutor strategies could alter their behavior based on knowledge tracing of the student log data. Also, new interface display applications are under consideration, using the interface module API. As mentioned, such interfaces could include Unreal Tournament^{TM}, Macromedia Flash^{TM}, or a Microsoft .NET application. We believe the customizable nature of the XTA could make it a valuable tool in the continued evolution of Intelligent Tutoring Systems.

2.3 The Assistment Builder

The foundation of the Assistment architecture is the content representation, an XML (eXensible Markup Language) schema that defines each problem. A problem consists of an interface definition and behavior definition. The interface definition provides a collection of simple widgets to be displayed to the student. The behavior definition is a representation of the state graph and its transitions, or a cognitive model (e.g. JESS rules). Many types of behaviors are possible within the representation and architecture. These two parts of the representation are consumed by the runtime Assistment architecture, and presented to the student over the web. Student actions are then fed back to the representation, and compared with the state graph or used to model trace.

2.3.1 Purpose of the Assistment Builder

The XML representation of content provides a base for which we can rapidly create specific pseudo-tutors. We sought to create a tool that would provide a simple web-based interface for creating these pseudo-tutors. Upon content creation, we could rapidly deploy the tutor across the web, and if errors were found with the tutor, bug-fixing or correction would be quick and simple. Finally, the tool had to be usable by someone with no programming experience and no ITS background. This applied directly to our project of creating tutors for the mathematics section of the Massachusetts Comprehensive Assessment System (MCAS) test [10]. We wanted the teachers in the public school system to be able to build pseudo-tutors. These pseudo-tutors are often referred to as Assistments, but the term is not limited to pseudo-tutors.

A secondary purpose of the Assistment Builder was to aid the construction of a Transfer Model. A Transfer Model is a cognitive model construct divorced from specific tutors. The Transfer Model is a directed graph of knowledge components representing specific concepts that a student could learn. These knowledge components are then associated with a specific tutor (or even sub-question within that tutor) so that the tutor is associated with a number of knowledge components. This allows us to maintain a complex cognitive model of the student without necessarily involving a production rule system. It also allows analysis of student performance in the context of the Transfer Model, resulting in knowledge tracing [2] and other useful methods. The simplest way to "mark" tutors in a Transfer Model is to associate the tutors (or their sub-questions) with knowledge components when the tutors are created. The Transfer Model created by the Assistment team is used to classify 8th grade mathematics items and has approximately 90 knowledge components. Over the six months since the inception of the Assistment Builder, nearly 1000 individual problems have thus far been associated with these 90 knowledge components.

2.3.2 Assistments

The basic structure of an Assistment is a top-level question that can then branch to scaffolding questions based on student input. The Assistment Builder interface uses only a subset of the full content XML representation, with the goal of producing simple pseudo-tutors. Instead of allowing arbitrary construction of question interfaces there are only five widget choices available to a content creator. These are radio-buttons, pull-down menus, checkboxes, text-fields, and algebra text fields that automatically evaluate mathematical expressions. The state graphs for each question are limited to two possible states. An arc occurs between the two states when the end-user answers a question properly. The student will remain in the initial state until the question is answered properly or skipped programmatically.

The scaffolding questions mentioned above are all queued as soon as a user gets the top-level question incorrect, or requests help in the form of a hint (for either event, the top-level question is skipped). Upon successfully completing the displayed scaffolding question the next is displayed until the queue is empty. Once the queue is empty, the problem is considered complete. This type of linear Assistment can be easily made with our tool, by first creating the main item and then the subsequent scaffolding questions. When building an Assistment a user may also add questions that will appear when a specific incorrect answer is received. This allows branches to be built that tutor along a "line of reasoning" in a problem, which adds more generality than a simple linear Assistment. Many Assistment authors also use text feedback on certain incorrect answers. These feedback messages are called buggy messages. Buggy messages address the specific error made, and match common or expected mistakes.

Content creators can also use the Assistment Builder to add hint messages to problems, providing the student with hints attached to a specific scaffolding question. This combination of hints, buggy messages, and branched scaffolding questions allow even the simple state diagrams described above to assume a useful complexity. Assistments constructed with the Assistment Builder can provide a tree of scaffolding questions branched from a main question. Each question consists of a customized interface, hint messages and bug messages, along with possible further branches.

2.3.3 Web Deployment

We constructed the Assistment Builder as a web application for accessibility and ease of use. A teacher or content creator can create, test, and deploy an Assistment without installing any additional software. Users can design and test their Assistments and then instantly deploy them. If further changes or editing are needed the Assistment can be loaded into the builder, modified, and then redeployed across all the curriculums that make use of the tutor. By making the Assistment Builder available over the web, there is no need for users to update any software if a new feature is added. They reap the benefits of any changes made to the system as soon as they log on. By storing created Assistments locally on our servers, allowing end-users to easily create a curriculum and assign it to a class for use by students is a simple task.

2.3.4 Features

Though there are many significant improvements to be made to the Assistment Builder's user interface, it is usable and reasonably easy to learn. When users first begin to use the Assistment Builder they will be greeted by the standard blank skeleton question. The initial blank skeleton question will be used to create the root question. The user will enter the question text, images, answers, and hint messages to complete the root question. After these steps the appropriate scaffolding is added. The question layout is separated into several views the Main View, All Answer View, Correct Answer View, Incorrect Answer View, Hints View, and Transfer Model View. Together these views allow users to highly customize their questions and their subsequent scaffolding.

Initially the user is presented with the Main View (see Figure 4). In this view question text, correct answers, and images can be added to the question. Additionally the user can add new scaffolding off of the current question, and specify if they would like the answers to be in a sorted or random order. The Main View is designed to gather the absolute minimum information needed to generate a question.

Upon completion of the items in the Main View the user then has the option to move to other views in order to further refine the question. Typically the next view to complete is the All Answer View. In the All Answers View the user has the option to add additional correct answers as well as incorrect answers. The incorrect answers serve two purposes. First, they allow a teacher to specify the answers students are likely to choose incorrectly and provide feedback in the form of a message or scaffolding. Second, the user can populate a list of answers for multiple choice questions. The user now has the

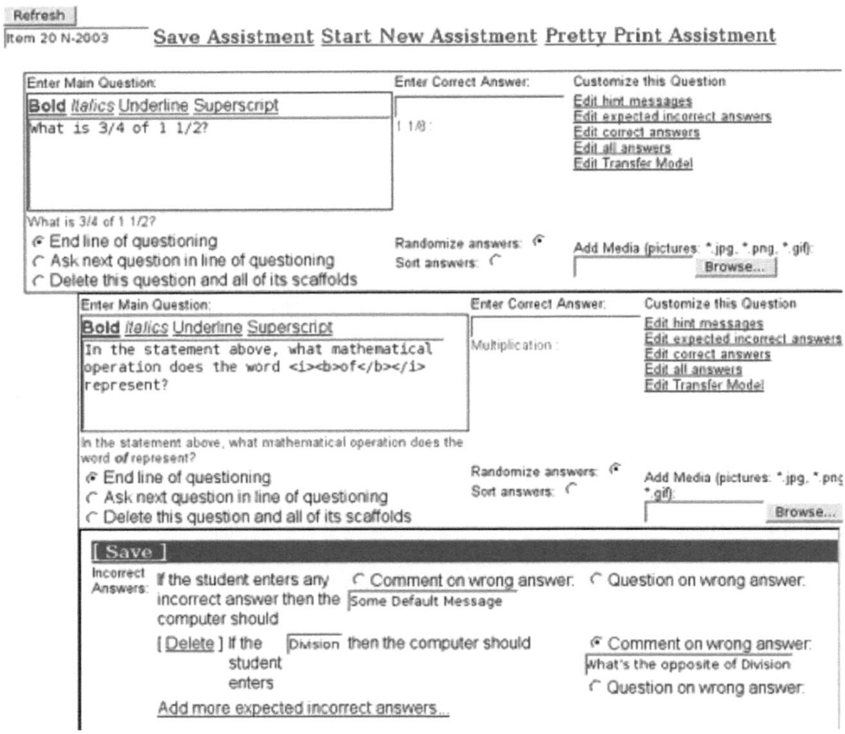

Fig. 2.4. The Assistment builder - initial question, one scaffold and incorrect answer view.

option to specify a message to be displayed for an incorrect answer or the option to scaffold. If the scaffolding option is chosen a new question block will appear indented below the current question. In the Hints View messages can be created that will be presented to the student when a hint is requested. Hints can consist of text, an image, or both. Multiple hint messages can be entered; one message will appear per hint request in the order that they are listed in this view. The final view is the Transfer Model View (see Figure 5). In this view the user has the option of specifying one or more skills that are associated with this particular question.

As mentioned above there are two methods of providing scaffolding questions: either by selecting Ask Next Line of Questioning from the Main View

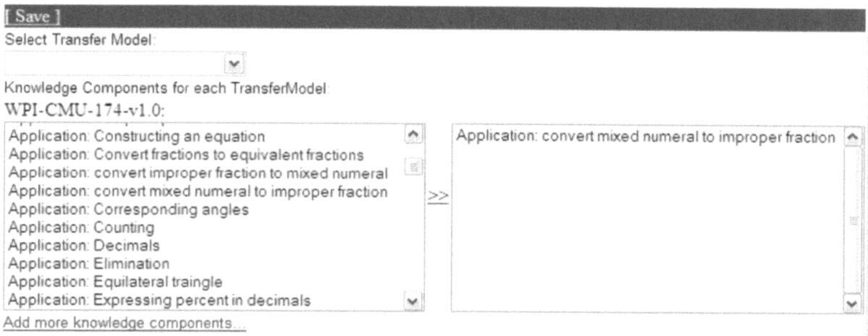

Fig. 2.5. Transfer model view.

or specify scaffolding on a specific incorrect answer. In utilizing either of these methods a new skeleton question will be inserted into the correct position below the current question. Creating a scaffolding question is done in the exact manner as the root question. After saving the Assistment the tutor is ready to be used. An Assistment can be modified at any time by loading it into the Assistment Builder and changing its properties accordingly. A completed running Assistment can be seen in Figure 6.

2.3.5 Methods

To analyze the effectiveness of the Assistment Builder, we developed a system to log the actions of an author. While authors have been constructing items for nearly six months, only very recently has the Assistment Builder had the capability to log actions.

Each action is recorded with associated meta-data, including author, timestamps, the specific series of problems being worked on, and data specific to each action. These actions were recorded for a number of Assistment authors over several days. The authors were asked to build original items and keep track of roughly how much time spent on each item for corroboration. The authors were also asked to create "morphs," a term used to indicate a new problem that had a very similar setup to an existing problem. "Morphs" are usually constructed by loading the existing problem into the Assistment Builder, altering it, and saving it with a different name. This allows rapid content development for testing transfer between problems. We wanted to compare the development time for original items to that of "morphs" [10]. To test the usability of the Assistment Builder, we were able to provide the software to two high-school teachers in the Worcester, Massachusetts area. These teachers were computer literate, but had no previous experience with intelligent tutoring systems, or creating mathematics educational software. Our tutorial consisted of demonstrating the creation of a problem using the Assistment Builder, then allowing the teachers to create their own with an

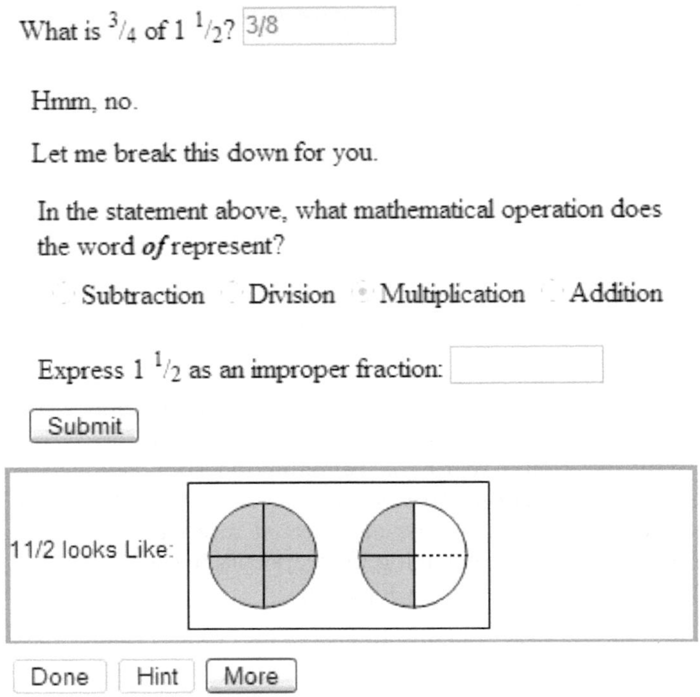

Fig. 2.6. An Assistment running.

experienced observer to answer questions. Finally, we allowed them to author Assistments on their own, without assistance.

2.3.6 Results and analysis

Prior to the implementation of logging within the Assistment Builder, we obtained encouraging anecdotal results of the software's use. A high-school mathematics teacher was able to create 15 items and morph each one, resulting in 30 Assistments over several months. Her training consisted of approximately four hours spread over two days in which she created 5 original Assistments under supervision. While there is unfortunately no log data to strengthen this result, it is nonetheless encouraging.

The logging data obtained suggests that the average time to build an entirely new Assistment is approximately 25 minutes. Entirely new Assistments are those that are built using new content and not based on existing material. This data was acquired by examining the time that elapsed between the initialization of a new problem and when it was saved. Creation times for Assistments with more scaffolds naturally took longer than those with fewer scaffolds. Experience with the system also decreases Assistment creation time, as end-users who are more comfortable with the Assistment Builder are able to work faster. Nonetheless, even users who were just learning the system were able to create Assistments in reasonable time. For instance, Users 2, 3, and 4 (see Table 1) provide examples of end-users who have little experience using the Assistment Builder. In fact, some of them are using the system for the first time in the examples provided.

Table 2.1. Full Item Creation

Username	Number of scaffolds	Time elapsed (min)
User1	10	35
User1	2	23
User2	3	45
User2	2	31
User2	0	8
User3	2	21
User4	3	37
User4	0	15
User5	4	30
User5	2	8
User5	4	13
User5	4	35
User5	3	31
User5	2	24
		Average: 25.4 minutes

We were also able to collect useful data on morph creation time and Assistment editing time. On average morphing an Assistment takes approximately 10-20 minutes depending on the number of scaffolds in an Assistment and the nature of the morph. More complex Assistment morphs require more time because larger parts of an Assistment must be changed. Editing tasks usually involve minor changes to an Assistment's wording or interface. These usually take less than a minute to locate and fix.

2.3.7 Future Work

In our continuing efforts to provide a tool that is accessible to even the most novice users we are currently working on two significant enhancements to the Assistment Builder. The first enhancement is a simplified interface that is both user-friendly and still provides the means to create powerful scaffolding pseudo-tutors. The most significant change to the current interface is the addition of a tab system that will allow the user to clearly navigate the different components of a question. The use of tabs allows us to present the user with only the information related to the current view, reducing the confusion that sometimes takes place in the current interface.

The second significant enhancement is a new question type. This question type will allow a user to create a question with multiple inputs of varying type. The user will also be able to include images and Macromedia Flash movies. Aside from allowing multiple answers in a single question, the new question type allows a much more customizable interface for the question. Users can add, in any order, a text component, a media component, or an answer component. The ability to place a component in any position in the question will allow for a more "fill in the blank" feel for the question and provide a more natural layout. This new flexibility will no longer force questions into the text, image, answer format that is currently used.

2.4 Content Development and Usage

In December of 2003, we met with the Superintendent of the Worcester Public Schools in Massachusetts, and were subsequently introduced to the three math department heads of 3 out of 4 Worcester middle schools. The goal was to get these teachers involved in the design process of the Assistment System at an early stage. The main activity done with these teachers was meeting about one hour a week to do "knowledge elicitation" interviews, whereby the teachers helped design the pedagogical content of the Assistment System.

2.4.1 Content Development

The procedure for knowledge elicitation interviews went as follows. A teacher was shown a Massachusetts Comprehensive Assessment System (MCAS) test item and asked how she would tutor a student in solving the problem. What kinds of questions would she ask the student? What hints would she give? What kinds of errors did she expect and what would she say when a student made an expected error? These interviews were videotaped and the interviewer took the videotape and filled out an "Assistment design form" from the knowledge gleaned from the teacher. The Assistment was then implemented using the design form. The first draft of the Assistment was shown to the teacher to get her opinion and she was asked to edit it. Review sessions with the teachers were also videotaped and the design form revised as needed. When the teacher was satisfied, the Assistment was released for use by students. For instance, a teacher was shown a MCAS item on which her students did poorly, such as item number 19 from the year 2003, which is shown in Figure 7. About 15 hours of knowledge elicitation interviews were used to help guide the design of Assistments.

Figure 8 shows an Assistment that was built for item 19 of 2003 shown above. Each Assistment consists of an original item and a list of scaffolding questions (in this case, 5 scaffolding questions). The first scaffolding question

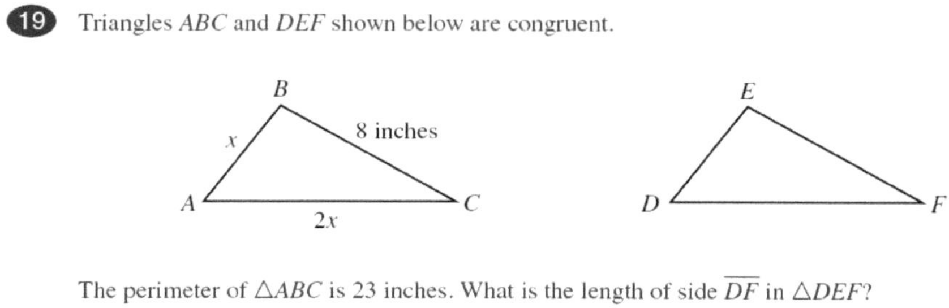

Fig. 2.7. Item 19 from the 2003 MCAS.

appears only if the student gets the item wrong. Figure 8 shows that the student typed "23" (which happened to be the most common wrong answer for this item from the data collected). After an error, students are not allowed to try the item further, but instead must then answer a sequence of scaffolding questions (or "scaffolds") presented one at a time.[5] Students work through the scaffolding questions, possibly with hints, until they eventually get the problem correct. If the student presses the hint button while on the first scaffold, the first hint is displayed, which would have been the definition of congruence in this example. If the student hits the hint button again, the hint describes how to apply congruence to this problem. If the student asks for another hint, the answer is given. Once the student gets the first scaffolding question correct (by typing AC), the second scaffolding question appears.

If the student selected 1/2 * 8x in the second scaffolding question, a buggy message would appear suggesting that it is not necessary to calculate area. (Hints appear on demand, while buggy messages are responses to a particular student error). Once the student gets the second question correct, the third appears, and so on. Figure 8 shows the state of the interface when the student is done with the problem as well as a buggy message and two hints for the 4th scaffolding question.

About 200 students used the system in May 2004 in three different schools from about 13 different classrooms. The average length of time was one class period per student. The teachers seemed to think highly of the system and, in particular, liked that real MCAS items were used and that students received instructional assistance in the form of scaffolding questions. Teachers also like that they can get online reports on students' progress from the Assistment web site and can even do so while students are using the Assistment System in their classrooms. The system has separate reports to answer the following questions about items, student, skills and student actions: Which items are my students

[5] As future work, once a predictive model has been built and is able to reliably detect students trying to "game the system" (e.g., just clicking on answer) students may be allowed to re-try a question if they do not seem to be "gaming". Thus, studious students may be given more flexibility.

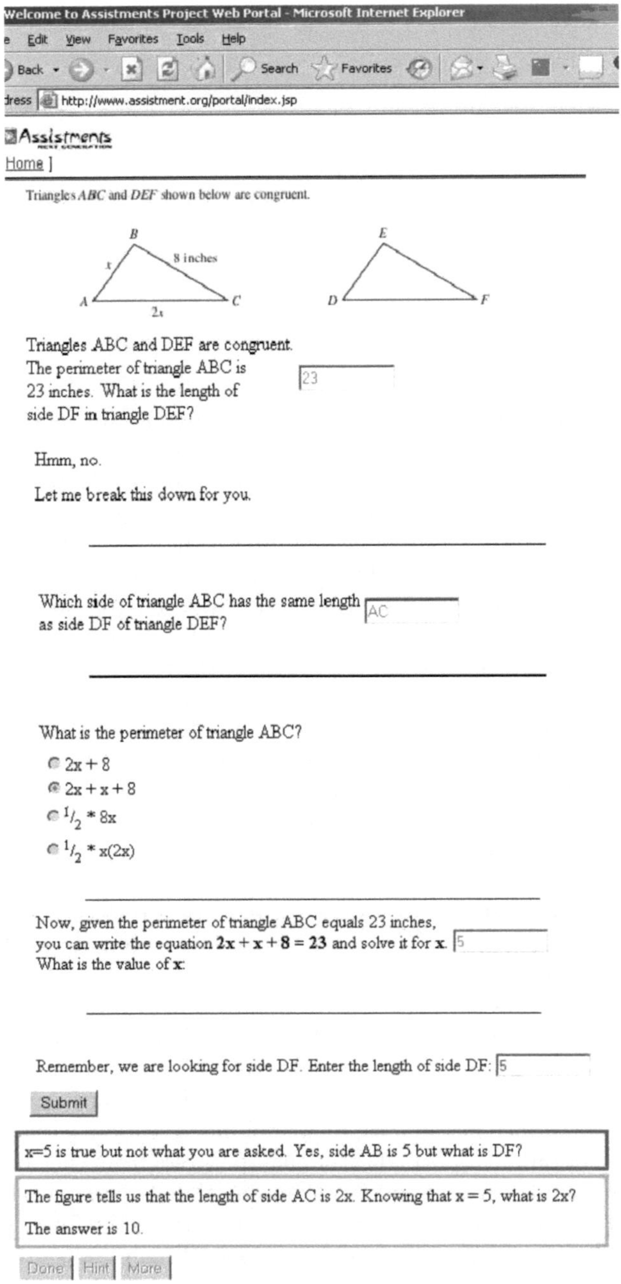

Fig. 2.8. An Assistment show just before the student hits the "done" button, showing two different hints and one buggy message that can occur at different points.

finding difficult? Which items are my students doing worse on compared to the state average? Which students are 1) doing the best, 2) spending the most time, 3) asking for the most hints etc.? Which of the approximately 90 skills that we are tracking are students doing the best/worst on? What are the exact actions that a given student took?

The three teachers from this first use of the Assistment System were impressed enough to request that all the teachers in their schools be able to use the system the following year. Currently that means that about 1,000 students are using the system for about 20 minutes per week for the 2004-2005 school year. Two schools have been using the Assistment System since September. A key feature of the strategy for both teacher recruitment and training is to get teachers involved early in helping design Assistments through knowledge elicitation and feedback on items that are used by their students.

We have spent considerable time observing its use in classrooms; for instance, one of the authors has logged over 50 days, and was present at over 300 classroom periods. This time is used to work with teachers to try to improve content and to work with students to note any misunderstandings they sometimes bring to the items. For instance, if it is noted that several students are making similar errors that were not anticipated, the Assistment Builder can be logged into and a buggy message added that addresses the students' misconception.

2.4.2 Database Reporting

The Assistment system produces reports individually for each teacher. These reports can inform the teacher about 1) "Which of the 90 skills being tracked are the hardest? 2) Which of the problems are students doing the poorest at and 3) reports about individual students. Figure 9 shows the "Grade book" report that shows for each student the amount of time spent in the system, the number of items they did, and their total score. Teachers can click on refresh and get instant updates. One of the common uses of this report is to track how many hints each student is asking for. We see that "Mary" has received a total of 700 over the course of 4 hours using the system, which suggests to teachers Mary might be using the system's help too much, but at this point it is hard to tell, given that Mary is doing poorly already.

2.4.3 Analysis of data to determine whether the system reliably predicts MCAS performance

One objective the project had was to analyze data to determine whether and how the Assistment System can predict students' MCAS performance. In Bryant, Brown and Campione [2], they compared traditional testing paradigms against a dynamic testing paradigm. In the dynamic testing paradigm a student would be presented with an item and when the student appeared to not be making progress, would be given a prewritten hint. If the

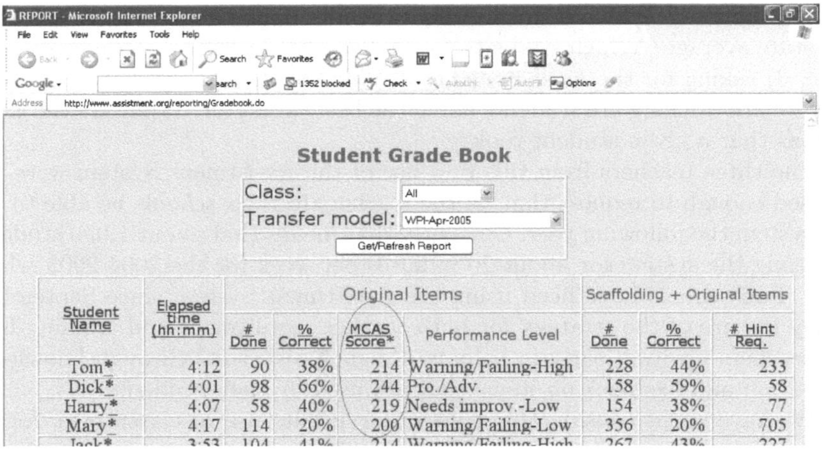

Fig. 2.9. The grade book report.

student was still not making progress, another prewritten hint was presented and the process was repeated. In this study they wanted to predict learning gains between pretest and posttest. They found that static testing was not as well correlated (R = 0.45) as with their "dynamic testing" (R = 0.60).

Given the short use of the system in May, 2004, there was an opportunity to make a first pass at collecting such data. The goal was to evaluate how well on-line use of the Assistment System, in this case for only about 45 minutes, could predict students' scores on a 10-item post-test of selected MCAS items. There were 39 students who had taken the post-test. The paper and pencil post-test correlated the most with MCAS scores with an R-value of 0.75.

A number of different metrics were compared for measuring student knowledge during Assistment use. The key contrast of interest is between a static metric that mimics paper practice tests by scoring students as either correct or incorrect on each item, with a dynamic assessment metric that measures the amount of assistance students need before they get an item correct. MCAS scores for 64 of the students who had log files in the system were available. In this data set, the static measure did correlate with the MCAS, with an R-value of 0.71 and the dynamic assistance measure correlates with an R-value of -0.6. Thus, there is some preliminary evidence that the Assistment System may predict student performance on paper-based MCAS items.

It is suspected that a better job of predicting MCAS scores could be done if students could be encouraged to take the system seriously and reduce "gaming behavior". One way to reduce gaming is to detect it [1] and then to notify the teacher's reporting session with evidence that the teacher can use to approach the student. It is assumed that teacher intervention will lead to reduced gaming behavior, and thereby more accurate assessment, and higher learning.

The project team has also been exploring metrics that make more specific use of the coding of items and scaffolding questions into knowledge components that indicate the concept or skill needed to perform the item or scaffold correctly. So far, this coding process has been found to be challenging, for instance, one early attempt showed low inter-rater reliability. Better and more efficient ways to use student data to help in the coding process are being sought out. It is believed that as more data is collected on a greater variety of Assistment items, with explicit item difficulty designs embedded, more data-driven coding of Assistments into knowledge components will be possible.

Tracking student learning over time is of interest, and assessment of students using the Assistment system was examined. Given that there were approximately 650 students using the system, with each student coming to the computer lab about 7 times, there was a table with 4550 rows, one row for each student for each day, with an average percent correct which itself is averaged over about 15 MCAS items done on a given day. In Figure 10, average student performance is plotted versus time. The y-axis is the average percent correct on the original item (student performance on the scaffolding questions is ignored in this analysis) in a given class. The x-axis represents time, where data is bunched together into months, so some students who came to the lab twice in a month will have their numbers averaged. The fact that most of the class trajectories are generally rising suggests that most classes are learning between months.

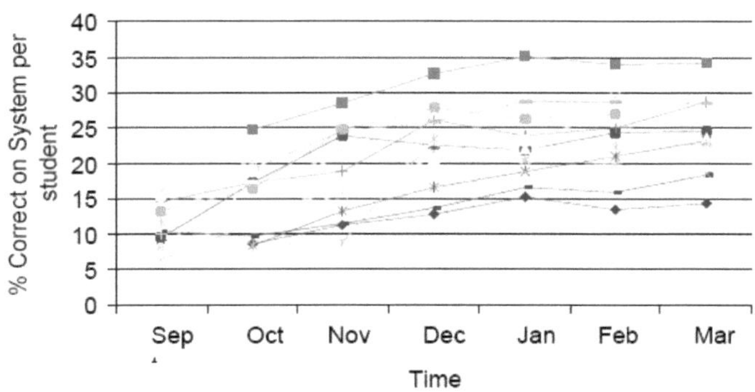

Fig. 2.10. Average student performance is plotted over time.

Given that this is the first year of the Assistment project, new content is created each month, which introduces a potential confounder of item difficulty. It could be that some very hard items were selected to give to students in September, and students are not really learning but are being tested on easier items. In the future, this confound will be eliminated by sampling items randomly. Adding automated applied longitudinal data analysis [7] is currently being pursued.

2.4.4 Analysis of data to determine whether the system effectively teaches.

The second form of data comes from within Assistment use. Students potentially saw 33 different problem pairs in random order. Each pair of Assistments included one based on an original MCAS item and a second "morph" intended to have different surface features, like different numbers, and the same deep features or knowledge requirements, like approximating square roots. Learning was assessed by comparing students' performance the first time they were given one of a pair with their performance when they were given the second of a pair. If students tend to perform better on the second of the pair, it indicates that they may have learned from the instructional assistance provided by the first of the pair.

To see that learning happened and generalized across students and items, both a student level analysis and an item level analysis were done. The hypothesis was that students were learning on pairs or triplets of items that tapped similar skills. The pairs or triplet of items that were chosen had been completed by at least 20 students.

For the student level analysis there were 742 students that fit the criteria to compare how students did on the first opportunity versus the second opportunity on a similar skill. A gain score per item was calculated for each student by subtracting the students' score (0 if they got the item wrong on their first attempt, and 1 if they got it correct) on their 1st opportunities from their scores on the 2nd opportunities. Then an average gain score for all of the sets of similar skills that they participated in was calculated. A student analysis was done on learning opportunity pairs seen on the same day by a student and the t-test showed statistically significant learning ($p = 0.0244$). It should be noted that there may be a selection effect in this experiment in that better students are more likely to do more problems in a day and therefore more likely to contribute to this analysis. An item analysis was also done. There were 33 different sets of skills that met the criteria for this analysis. The 5 sets of skills that involved the most students were: Approximating Square Roots (6.8% gain), Pythagorean Theorem (3.03% gain), Supplementary Angles and Traversals of Parallel Lines (1.5% gain), Perimeter and Area (Figure 11)(4.3% gain) and Probability (3.5% gain). A t-test was done to see if the average gain scores per item were significantly different than zero, and the result ($p = 0.3$) was not significant. However, it was noticed that there was a large number of negative average gains for items that had fewer students so the average gain scores were weighted by the number of students, and the t-test was redone. A statistically significant result ($p = 0.04$) suggested that learning should generalize across problems. The average gain score over all of the learning opportunity pairs is approximately 2%. These results should

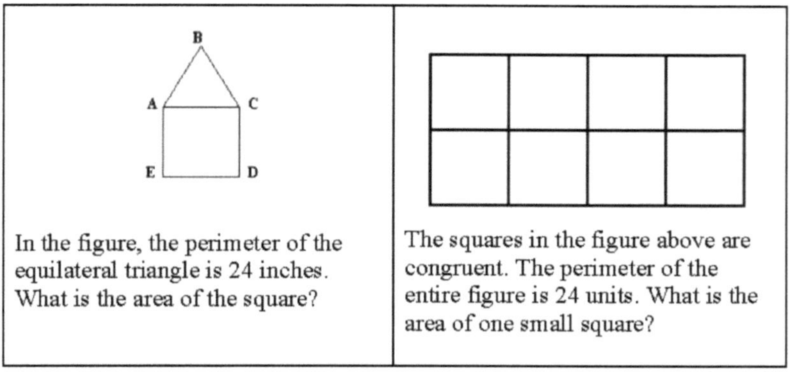

Fig. 2.11. A perimeter and area learning opportunity pair.

be interpreted with some caution as some of the learning opportunity pairs included items that had tutoring that may have been less effective. In fact, a few of the pairs had no scaffolding at all but just hints.

2.4.5 Experiments

The Assistment System allows randomized controlled experiments to be carried out. At present, there is control for the number of items presented to a student, but soon the system will be able to control for time, as well. Next, two different uses of this ability are described.

Do different scaffolding strategies affect learning?

The first experiment was designed as a simple test to compare two different tutoring strategies when dealing with proportional reasoning problems like item 26 from the 2003 MCAS: "The ratio of boys to girls in Meg's chorus is 3 to 4. If there are 20 girls in her chorus, how many boys are there?" One of the conditions of the experiment involved a student solving two problems like this with scaffolding that first coached them to set up a proportion. The second strategy coached students through the problem but did not use the formal notation of a proportion. The experimental design included two items to test transfer. The two types of analyses the project is interested in fully automating is 1) to run the appropriate ANOVA to see if there is a difference in performance on the transfer items by condition, and 2) to look for learning during the condition, and see if there is a disproportionate amount of learning by condition.

Two types of analyses were done. First, an analysis was done to see if there was learning during the conditions. 1st and 2nd opportunity was treated as a repeated measure and to look for a disproportionate rate of learning due to condition (SetupRatio vs. NoSetup). A main effect of learning between first and second opportunity ($p = 0.05$) overall was found, but the effect of condition was not statistically significant ($p = 0.34$). This might be due to the fact that the analysis also tries to predict the first opportunity when there is no reason to believe those should differ due to controlling condition assignment. Given that the data seems to suggest that the SetupRatio items showed learning a second analysis was done where a gain score (2nd opportunity minus 1st opportunity) was calculated for each student in the SetupRatio condition, and then a t-test was done to see if the gains were significantly different from zero and they were ($t = 2.5$, $p = 0.02$), but there was no such effect for NoSetup.

The second analysis done was to predict each student's average performance on the two transfer items, but the ANOVA found that even though the SetupRatio students had an average score of 40% vs. 30%, this was not a statistically significant effect.

In conclusion, evidence was found that these two different scaffolding strategies seem to have different rates of learning. However, the fact that setting up a proportion seems better is not the point. The point is that it is a future goal for the Assistment web site to do this sort of analysis automatically for teachers. If teachers think they have a better way to scaffold some content, the web site should send them an email as soon as it is known if their method is better or not. If it is, that method should be adopted as part of a "gold" standard.

Are scaffolding questions useful compared to just hints on the original question?

An experiment was set up where students were given 11 probability items. In the first condition, the computer broke each item down into 2-4 steps (or scaffolds) if a student got the original item wrong. In the other condition, if a student made an error they just got hints upon demand. The number of items was controlled for. When students completed all 11 items, they saw a few items that were morphs to test if they could do "close"-transfer problems.

The results of the statistical analysis were showing a large gain for those students that got the scaffolding questions, but it was discovered that there was a selection-bias. There were about 20% less students in the scaffolding condition that finished the curriculum, and those students that finished were probably the better students, thus making the results suspect. This selection bias was possible due to a peculiarity of the system that presents a list of assignments to students. The students are asked to do the assignments in order, but many students choose not to, thus introducing this bias. This will be easy to correct by forcing students to finish a curriculum once they have started it. For future work, a new experiment to answer this question, as well as several other questions, will be designed and analyzed.

2.4.6 Survey of students' attitudes

At the end of the 2004-2005 school year, the students using the Assistment system participated in a survey. 324 students participated in the survey and they were asked to rate their attitudes on statements by choosing Strongly Agree, Agree, Neither Agree nor Disagree, Disagree or Strongly Disagree. The students were presented with statements such as "I tried to get through difficult problems as quickly as possible," and "I found many of the items frustrating because they were too hard." The statements addressed opinions about subjects such the Assistment system, math, and using the computer.

We wanted to find out what survey questions were correlated with initial percent correct and learning in the Assistment system. The responses to "I tried to get through difficult problems as quickly as possible," were negatively correlated with learning in the Assistment system ($p = -0.122$). The responses to "When I grow up I think I will use math in my job," were positively correlated with learning in the Assistment system ($p = 0.131$). Responses to statements such as "I am good at math," "I work hard at math," and "I like math class," were all positively correlated with students' percent correct in September (at the beginning of Assistment participation).

We believe that the survey results point to the importance of student motivation and attitude in mastering mathematics. For future work, we plan to examine ways to increase student motivation and keep them on task when working on Assistments.

2.5 Summary

The Assistment System was launched and presently has 6 middle schools using the system with all of their 8th grade students. Some initial evidence was collected that the online system might do a better job of predicting student knowledge because items can be broken down into finer grained knowledge components. Promising evidence was also found that students were learning during their use of the Assistment System. In the near future, the Assistment project team is planning to release the system statewide in Massachusetts.

References

1. Anderson, J. R. (1993). Rules of the mind. Hillsdale, NJ: Erlbaum.
2. Anderson, J. R., Corbett, A. T., Koedinger, K. R., and Pelletier, R. (1995). Cognitive tutors: Lessons learned. The Journal of the Learning Sciences, 4 (2), 167-207.
3. Anderson, J.R., and Pelletier, R. (1991). A development system for model-tracing tutors. In Proceedings of the International Conference of the Learning Sciences, 1-8.

4. Baker, R.S., Corbett, A.T., Koedinger, K.R. (2004) Detecting Student Misuse of Intelligent Tutoring Systems. Proceedings of the 7th International Conference on Intelligent Tutoring Systems, 531-540.
5. Campione, J.C., Brown, A.L., and Bryant, N.R. (1985). Individual differences in learning and memory. In R.J. Sternberg (Ed.). Human abilities: An information-processing approach, 103-126. New York: W.H. Freeman.
6. Feng, Mingyu, Heffernan, N.T. (2005). Informing Teachers Live about Student Learning: Reporting in the Assistment System. Submitted to the 12th Annual Conference on Artificial Intelligence in Education 2005, Amsterdam
7. Heffernan, N. T. and Croteau, E. (2004) Web-Based Evaluations Showing Differential Learning for Tutorial Strategies Employed by the Ms. Lindquist Tutor. Proceedings of 7th Annual Intelligent Tutoring Systems Conference, Maceio, Brazil. Pages 491-500.
8. Jackson, G.T., Person, N.K., and Graesser, A.C. (2004) Adaptive Tutorial Dialogue in AutoTutor. Proceedings of the workshop on Dialog-based Intelligent Tutoring Systems at the 7th International conference on Intelligent Tutoring Systems. Universidade Federal de Alagoas, Brazil, 9-13.
9. Jarvis, M., Nuzzo-Jones, G. and Heffernan. N. T. (2004) Applying Machine Learning Techniques to Rule Generation in Intelligent Tutoring Systems. Proceedings of 7th Annual Intelligent Tutoring Systems Conference, Maceio, Brazil. Pages 541-553
10. Koedinger, K. R., Aleven, V., Heffernan. T., McLaren, B. and Hockenberry, M. (2004) Opening the Door to Non-Programmers: Authoring Intelligent Tutor Behavior by Demonstration. Proceedings of 7th Annual Intelligent Tutoring Systems Conference, Maceio, Brazil. Pages 162-173
11. Koedinger, K. R., Anderson, J. R., Hadley, W. H., and Mark, M. A. (1997). Intelligent tutoring goes to school in the big city. International Journal of Artificial Intelligence in Education, 8, 30-43.
12. Livak, T., Heffernan, N. T., Moyer, D. (2004) Using Cognitive Models for Computer Generated Forces and Human Tutoring. 13th Annual Conference on (BRIMS) Behavior Representation in Modeling and Simulation. Simulation Interoperability Standards Organization. Arlington, VA. Summer 2004
13. Mitrovic, A., and Ohlsson, S. (1999) Evaluation of a Constraint-Based Tutor for a Database Language. Int. J. on Artificial Intelligence in Education 10 (3-4), pp. 238-256.
14. Mostow, J., Beck, J., Chalasani, R., Cuneo, A., and Jia, P. (2002c, October 14-16). Viewing and Analyzing Multimodal Human-computer Tutorial Dialogue: A Database Approach. Proceedings of the Fourth IEEE International Conference on Multimodal Interfaces (ICMI 2002), Pittsburgh, PA, 129-134.
15. Murray, T. (1999). Authoring intelligent tutoring systems: An analysis of the state of the art. International Journal of Artificial Intelligence in Education, 10, pp. 98-129.
16. Nuzzo-Jones, G., Walonoski, J.A., Heffernan, N.T., Livak, T. (2005). The eXtensible Tutor Architecture: A New Foundation for ITS. In C.K. Looi, G. McCalla, B. Bredeweg, and J. Breuker (Eds.) Proceedings of the 12th Artificial Intelligence In Education, 902-904. Amsterdam: ISO Press.
17. Razzaq, L., Feng, M., Nuzzo-Jones, G., Heffernan, N.T., Koedinger, K. R., Junker, B., Ritter, S., Knight, A., Aniszczyk, C., Choksey, S., Livak, T., Mercado, E., Turner, T.E., Upalekar. R, Walonoski, J.A., Macasek. M.A.,

Rasmussen, K.P. (2005). The Assistment Project: Blending Assessment and Assisting. In C.K. Looi, G. McCalla, B. Bredeweg, and J. Breuker (Eds.) Proceedings of the 12th Artificial Intelligence In Education, 555-562. Amsterdam: ISO Press.

18. Rose, C. P. Gaydos, , A., Hall, B. S., Roque, A., K. VanLehn, (2003), Overcoming the Knowledge Engineering Bottleneck for Understanding Student Language Input , Proceedings of AI in Education.
19. Singer, J. D. and Willett, J. B. (2003). Applied Longitudinal Data Analysis: Modeling Change and Occurrence. Oxford University Press, New York.
20. Turner, T.E., Macasek, M.A., Nuzzo-Jones, G., Heffernan, N.T, Koedinger, K. (2005). The Assistment Builder: A Rapid Development Tool for ITS. In C.K. Looi, G. McCalla, B. Bredeweg, and J. Breuker (Eds.) Proceedings of the 12th Artificial Intelligence In Education, 929-931. Amsterdam: ISO Press.

3

Alife in the Classrooms: an Integrative Learning Approach

Jordi Vallverdú

Universitat Autònoma de Barcelona jordi.vallverdu@uab.es

The purpose of this contribution is to apply Alife systems (Artificial Life) to the integrative learning of computation, biology, mathematics and scientific epistemology (methods, practices...) in the classroom.

Contemporary trends in Artificial Life and a deep interest in game theory are used to create several kinds of models, which are useful for day-to-day human practices, not just for scientific ones. Leisure activities such as computer games are an example of this.

Using L-Systems (an automaton designed by Aristid Lindenmayer in 1968 to model cell development), students learn about the fractal nature of the natural world, introducing themselves to programming and to the new paradigm of e-Science; a collaborative and computational way to perform scientific activity. Creating these Alife worlds, students are introduced to virtual instruments and can also create hypertextual research strategies (working together with distant students from other places or countries).

Our proposal fits well with contemporary theories about extended mind and human cognition, offering an easy and cheap computational way to learn e-Science (both contents and practices).

3.1 An Integrative Model of Learning

Contemporary science, and I have in mind the paradigmatical case of Biology, is a cross-field research activity. To be able to discover DNA double strings, Watson & Crick used ideas, tools and methods from several disciplines, such as Chemistry, Biology and Physics [1]. Fifty years later the Human Genome Project achieved the human genome sequence with the contribution of Computer Scientists (creating at the same time a new field: Bioinformatics), Statisticians or Mathematicians [2].

A recent report from the National Research Council, *Bio 2010* [3], recommends that undergraduate biology preparation become more interdisciplinary.

Jordi Vallverdú: *Alife in the Classrooms: an Integrative Learning Approach*, Studies in Computational Intelligence (SCI) **44**, 51–76 (2007)
www.springerlink.com

Modern science (and modern scientific fields) requires a breadth of skills that go well beyond the limited set of experiences that undergraduate students receive in their courses [4]–[5]. Powerful innovations, such as the digital revolution, have changed the ways in which science is practiced. Computers play a central role in the acquisition, storage, analysis, interpretation and visualization of scientific data, a kind of data that is increasing every day in quantity (in amounts of petabytes of data: the 'data tsunami', [6]).

Starting from 'information' we achieve 'knowledge' through the contribution of computational tools. Robert Logan [7], talks about the 'Knowledge Era' idea and the increasing understanding process from *data* to *information, knowledge* and *wisdom,* where 'data' are raw, unprocessed facts and/or figures, often obtained via the use of measurement instruments, 'information' is data that has been processed and structured, adding context and increased meaning, 'knowledge' is the ability to use information tactically and strategically to achieve specified objectives and, finally, 'wisdom', is the ability to select objectives that are consistent with and supportive of a general set of values, such as human values.

Our students are provided with initial information, along with electronic tools and heuristic rules to transform it into (integrative) knowledge. And this is a practical project, in which there is a continous feedback relationship between teacher and learners, at the same time that learners are learning-by-doing. The imaging software, with the cognitive and aesthetic values implied in it, is a functional way to transform abstract ideas into 'real' things (considering visualizations as true model representations of the real world). With these images, we create human mental and physical landscapes, designing tools to make the high levels of abstraction required by contemporary scientific knowledge easier.

We can affirm that the construction of sense from huge amounts of raw data requires an increasing use of computational devices which enables a better cognitive framework. Imaging or visualization techniques are an example of this, and are usually called *SciVis*('Scientific Visualization'). At the same time, the graphical representation of complex scientific concepts can enhance both science and technology education. Now that scientific visualization programs can be used on the kinds of computers available in schools, it is feasible for teachers to make use of these tools in their science and technology education classes.

According to the NC State University College of Education - Graphic Communications Program (http://www.ncsu.edu/scivis/), a SciVis approach creates a curriculum which allows for:

- Enriching the curriculum by incorporating computer and network technology.
- Understanding the theoretical basis of using graphics to convey scientific information.
- Exploring how science and technology curricula can be integrated.
- Experiencing the emerging field of scientific visualization.

- Developing instructional materials for integrating scientific visualization
- into the science/technology classroom.
- Building working graphics/science/technology teams.

Image processing facilitates the manipulation of data in ways that revolutionize how information is perceived, analyzed, communicated, and stored. Image processing is now an important tool in many major areas of scientific study [8].

With these tools, students transform information data into scientific knowledge, through an inquiry-based approach. The inquiry-based approach has shifted the focus of science education from traditional memorization of facts and concepts in separate specific disciplines, to inquiry-based learning in which students are actively engaged using both science processes, and critical thinking skills as they search for answers [9]. A primary goal for current inquiry-based reforms in science education is that students develop an understanding of the nature of science by doing science [10]

Let's see an example: Hounshell and Hill [11] arranged to use computer simulations to expand, enrich, reconstruct, and supplement the laboratory and lecture components of the traditional biology course for students in Grades 12–13. Results indicated that computer simulations for selected laboratory, demonstration, and other classroom activities can make a difference in improving both attitudes and achievement of students enrolled in biology [12].

This is the reason for the necessity of active and integrative learning, integrating and interpreting knowledge from different disciplines, such as biology, informatics or mathematics [13]. Integration can be understood as 'bringing together information from different sources and structures'. But when information comes from several different sources and we obtain different knowledge from them, we need an integrative process of knowledge integration or blending, that is, the integration of different domains of knowledge into a single domain. And a blend of two domains won't consist of their sum or juxtaposition. Instead, this new domain will have its own structure and semantics, which, from one point of view, brings problems of interpretation and validation of emergent concepts, but from another represents a promising space for the generation of new ideas and solutions. To be able to do so, it is essential to have some sort of unifying process, as suggested by the Conceptual Blending Theory [14] or the Scaffolded Knowledge Integration framework [15]. The idea is that the intent of Knowledge Integration is to help students develop an integrated scientific understanding, linking isolated scientific concepts to each other and to the world outside the science classroom. We try to encourage students to explore the ways in which knowledge can be discussed and created from very different areas.

When we introduce our students to the use of LSE, they must integrate several kinds of knowledge (biology, mathematics, programming), cognitive skills (through visualization techniques) and to develop a mental framework for the new aspects of the geometrical nature of plants.

At the same time we can ask ourselves, as teachers and researchers of scientific truth and its communication and transmission, if, as Galileo said, "the book of nature is written in mathematical characters". What is the true reality of the world: the world or its (mathematical) models? But, perhaps, the true question is another one: can we think of the world without our (mathematical) models? And, when we are talking of the world, are we talking about the real world or of our models of the world? Our ideas about the nature of the world are provided by our models about the world. Therefore, models are the 'cultural glasses' by means of which we 'see' the world. So, virtual simulations are models of the world, and good simulations are, at some point, our best way of relating to the world.

If the basic nature and goals of scientific research have changed, why shouldn't we change our educational models? An integrative approach enables better knowledge, at the same time it requires that the specific knowledge involved in the whole process must be clearly understood. If the boundaries between disciplines are becoming arbitrary, the rational solution to that new situation should be to allow students to learn the different languages of the disciplines in context.

Besides, we must consider deep changes in contemporary science, that is, the transition to an *e-Science* with a new kind of knowledge production [16]. We live in a Network Society [17], with a networked science. The deepest change of contemporary science concerns computer empowerment in scientific practices. So e-Science is computationally intensive science carried out in highly distributed network environments, using huge data sets that require intensive computing (grids, clusters, supercomputers, distributed computing) [18].

¿From an integrative point of view I propose using L-systems (Lindenmayer systems) to put together several strategies:

1. **Knowledge strategies:** integrating biology, mathematics and computing (artificial life and programming) fields.
2. **Procedural strategies:** dynamics of e-Science (virtual experiments, databases, decentralized work, quantification and modeling of knowledge, open source software...).
3. **Cognitive strategies:** imaging and visualization of scientific information, considering the importance of emotional, aesthetic and cognitive aspects of learning. Representation by simulation is a successful way of thinking better.

All these strategies are embedded in L-systems, and it is necessary that teachers have these different approaches in mind. It would also be a good idea to develop cross-teaching with teachers of several subjects implied in the optimal development of L-systems.

3.2 An Educational Application of Cognitive Sciences

There are also more elements to consider in integrative models: the cognitive aspects of human reasoning and specifically, the student's ability to learn science. Although several authors talk about extended mind and computational extensions of the human body [19]–[22], most of these proposals don't analyze the deep epistemological implications of computer empowerment in scientific practices. They talk about new human physical and mental environments, not about new ways of reasoning in the broader sense of the term.

At the same time, we must identify the principal concept of e-Science: *Information*. Sociologists like Castells [17] or philosophers like Floridi [23] talk respectively about the Network Society with a 'culture of real virtuality', an open space sustained by the Information Technology (IT) revolution (and changes inside capitalist economic models and the pressure of new cultural and social movements), or a new space for thinking and debating: the *infosphere* [23]. We could also talk about a Philosophy of Information [24]–[25]. We must admit that despite the fact that there have been several philosophers who have tried to show the radical implications of computation in human reasoning, [26]–[30] it hasn't implied the design of a new epistemology for e-Science.

So, if *information* obtained by computational tools is the key of new e-Science, it is absolutely necessary to think about the ways we can produce, learn and communicate that information. For that purpose ideas from cognitive sciences are very useful, especially, those of the 'extended mind'.

Cognitive sciences have been increasingly invoked in discussions of teaching and learning [31]–[32], making emphasis on *metacognition* [33]–[34], that is, knowing what one knows and does not know, predicting outcomes, planning ahead, efficiently apportioning time and cognitive resources, and monitoring one's efforts to solve a problem or to learn. So, metacognition can be considered as the process of considering and regulating one's own learning, and potentially revising beliefs on the subject. Here, (1) learners do not passively receive knowledge but rather actively build (construct) it; (2) to understand something is to know relationships; (3) all learning depends on prior knowledge; and (4) successful problem-solving requires a substantial amount of qualitative reasoning [35]. So, metacognition is not directly related to a specific kind of heuristic coordinated with computational environments, but acquires it's nature as a whole process of active meaning creation.

Due to the abstract complexity of several fields of contemporary science and scientific knowledge, the learning tools have evolved in a way which uses virtual modelling. It is now commonly accepted that research on Intelligent Tutoring Systems (ITS), also sometimes called Intelligent Computer Aided Instructional (ICAI) systems, started as a distinct approach with a dissertation by Carbonell [36] and with his system SCHOLAR. ¿From this beginning, this research has developed in many directions, but broadly speaking, two major schools of thought have evolved and produced two different types of system:

learning environments and active teaching systems, as we can see in [37]:33 with Figure 3.1:

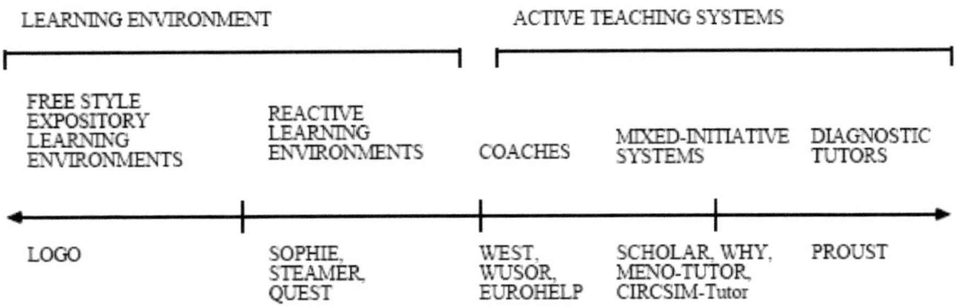

Fig. 3.1. Computer education tools

My approach is centered on creating a new learning environment through the possibilities of Alife. This is a coaching system in which the teacher helps the students to perform better and to integrate several kinds of knowledge under a common conceptual platform.

But I am talking about a *pragmatic* approach to knowledge integration. There is a strong relationship between *doing and thinking* [38]. And it is precisely in the process of 'doing' that contemporary science embraces computational tools. It is now, when we must incorporate the cognitive results from the extended mind model [39]. This advocates a special sort of externalism, an active externalism, based on the active role of the environment in driving cognitive processes. Environmental supports develop a crucial role in knowledge production. We are extending ourselves with these new instruments, consequently we should understand how the apparatus operates, and include its results in our own scientific abilities. We know and think with the helpful contribution of machines, so we should include them in our cognitive models. They are not only the instrument through which we achieve results from nature, but they are part of our minds and design the shape of our thoughts. Imaging computational tools are an example of that idea; 3-D visualizations have enabled profound progress in the scientific use of vast amounts of difficult data. User-friendly interfaces help us to make better representations of the world, and we must remember that approximately 60% of the human brain's sensory input comes from vision [12]. Images and animations are valuable tools in both producing and learning scientific topics, because they help users with important conceptual relationships [40]–[42].

We should consider another crucial aspect of contemporary theories of rationality: the inclusion of values inside those models. Emotions are something necessary to achieve rational processes and good results [43]–[45]. Pragmatic analyses of the roles of non-epistemic values in scientific enquiry probes the underlying existence of beliefs and values in scientific practices, revealing an important interplay between the aesthetic, cognitive and affective processes [46]. We could also say that artists and scientists share intuitions about the world [47], and electronic artists use technological tools in which scientific principles are embedded. So, the structures intuited by artists and scientists also involve the patterns of process in nature [48]. "All mathematicians experience a genuine sense of aesthetics", said Henri Poincaré [49]. New algorithms generate new shapes, while computer imaging techniques are enabling new visualizations. These techniques are extensional ways of human information processing. They are in some way, our mind. Visual-spatial thinking has been an aspect of science overlooked by educators [50], although there are relevant arguments in favour of its importance in the scientific bibliography [51]–[58]. Images are coherent encodings of experience, and the computational ways by which they are produced in the scientific process, belong to extended human cognition. We know that successful visual resources support important cognitive processes, and recent surveys of computer science educators suggest a widespread belief that visualization technology positively impacts learning [59]. If visualizations are now a cornerstone of most scientific endeavors, they must be soundly based on an understanding of cognition, which is provided by cognitive psychologists and philosophers [60]. The merging of scientific fields with disciplines such as art, psychology, and technology can result in visualizations that are not only effective in communicating concepts, but are also easily interpreted by new students [61]. These interdisciplinary collaborations are important for visualizations of the particulate level of matter to be effective learning tools. We must also consider that the cognitive advantages of *visual models* and their ability to explicitly show, in a single unified view, the relationships between a large number of diverse elements, makes them an indispensable part of the knowledge integration process. The goal of knowledge integration is to enable an emergent level of intelligence in the face of scientific complexity. As the Idiagram project explains [62]: *knowledge integration* is the process of fitting our ideas – our theories of how-the-world-works – together into a coherent structure. That coherent structure, and the *process* of bringing knowledge together, has a number of critically important, yet under-appreciated, uses: (a) as we expand the scope of our thinking we may come across just the idea, or combination of ideas, that enables progress on the seemingly intractable problems we face; (b) as we reconcile conflicting ideas we can force into the open, hidden assumptions and logical inconsistencies (c) as we synthesize diverse perspectives we can clarify our thinking and highlight areas of (in)coherence, (dis)agreement, or (un)certainty; (d) as we connect ideas we can create a whole that is greater than the sum of its parts.

According to the above mentioned cognitive reasons and arguments I now wish to apply the integrative learning model to Alife.

3.3 Alife as an Unified Scientific Enterprise

In 1987, Chris Langton instigated the notion of "Artificial Life" (or "Alife"), at a workshop in Los Alamos, New Mexico [63]. He was inspired by John Von Neumann and his early work on self-reproducing machines in cellular automata. The First researchers in Alife were inspired by biological systems to produce computational simulations and analytical models of organisms or biological reproduction and group behaviors [64]–[65].

Alife was applied very early to computer games, with *Creatures* (1997), programmed by Steve Grand, who was nominated by *The Sunday Times* as one of the "Brains Behind 21st Century" and was awarded an OBE in the Millennium Honors List [66]. The International Society of Artificial Life [67], or ISAL, has an official journal *Artificial Life,* which is published by MIT Press. Alongside these approaches a synthetic biology has appeared, which considers life as a special kind of chemistry and is able to create computer models that simulate replication and evolution *in silico* [68].

Several philosophical approaches have tried to analyze this virtual biology [69]–[74] and its epistemic consequences, concluding that computational tools are valuable for *real* science and Alife simulations provide a modeling vocabulary capable of supporting genuine communication between theoretical and empirical biologists.

We must also consider the emotional and aesthetic aspects of cognition implied in Alife systems. Alife is the coherent combination of engineering, computation, mathematics, biology and art. According to Whitelaw [75], these characteristics of Alife provide a very useful way to learn science in an integrative and contemporary way.

Artificial life, or Alife, is an interdisciplinary science focused on artificial systems that mimic the properties of living systems. In the 1990's, new media artists began appropriating and adapting the techniques of Alife science to create Alife art [76]–[77].

Alife art responds to the increasing technological nature of living matter by creating works that seem to mutate, evolve, and respond with a life of their own. Pursuing Alife's promise of emergence, these artists produce not only artworks but generative and creative processes: here creation becomes metacreation. At the same time, we could argue about the ontological status of a simulation, because the credibility of digital computer simulations has always been a problem [78]. Is Alife a true simulation? And what is the epistemic value of simulations? From an historical point of view, the question of computer simulation credibility is a very old one and there are different possible standpoints on the status of simulation: they can be considered as genuine experiments, as an intermediate step between theory and experiment or as a tool.

¿From a pragmatical philosophical perspective, I propose this lema: "does it work?". Alife simulations, at least L-systems, reproduce and explain plant development. In our relationship with the world the way to obtain truths is mediated by our models, and we know that our models fit the world well when they show a similar behavior, an homogeneous nature. Science solves problems and explains the nature of these problems. The prehistory of individual plant simulation can be traced back to to the Ulam's digital computer simulations on branching patterns with cellular automata, at the beginning of the 1960's. Then Lindenmayer's work on substitution formal systems - the so-called L-systems -, which were first published in 1968, helped some biologists to accept such a formal computer modeling. In 1979, De Reffye produced and published through his Ph.D. thesis the first universal 3D simulation of botanical plants. He could simulate them, whatever their "architectural model" in the sense of the term proposed by the botanist Hallé. From a conceptual point of view, the new architectural vision, due to Hallé's work in the 1970's, enabled De Reffye to consider plants as discrete events generating discrete trees and not as chemical factories.

Then, the question is "what kind of existence does the scientist ascribe to the mathematical or logical equivalent he is using to model his phenomenon?". My point is: if the model fits reality well, it shares an important *essence* with the real world. So, it is *the real world*, at least at some levels of its reality. We don't ask all the possible questions of the real world, just the ones we are able to think of in a formal way. So, if our approach to the reality of the facts is limited by our questions, and the world is never all the possible world but just the *thinkable* world, then simulations (or good simulations) are true experiments. We must also admit that the best map of the world is the world itself. Then, to operate properly with the world we should use the whole world, something which is impossible. Consequently, we reduce parts of the world to simple models, which can be 'real' (one plant as an example of all plants) or 'virtual' (a L-system representation of a plant). The problem is not the nature of the model, but its capacity to represent the characteristics of the world that we try to know (and to learn/teach).

As a consequence, we can consider Alife simulations as true real life observations. They are as limited as are our own theoretical models. There is nothing special in virtual simulations which cannot enable us to use them as true models of the world. The only question is to be sure about the limits of our virtual model, in the same way that when we go to the laboratory and analize *a* plant (or a limited series of them), that one is not the whole species, just a specific model. Although that plant is *real* we use them in the laboratory as a model representation of all the same items in the world. Consequently, we can suppose that the rest of the plants manifest a similar structure and behavior, if our chosen model is a good one (it could be a special mutation, or a bad example,...). The virtual model reaches a different level of abstraction, but it is also a model. The crucial question is about the accuracy of the similitudes between the virtual model and the real world, not about the biological o digital nature of the studied object.

With that conceptual framework, I propose the use of L-systems to introduce students to Alife. L-systems can achieve several degrees of complexity, from geometrical reproduction of living systems to the simulation of life environments. In my view, I think it is better to first introduce students to the geometrical and programming aspects of L-systems to be able to create complex life with L-systems at a later stage. Consequently, I consider this chapter to be an introduction to the elementary basis of L-systems, which should be continued in future writing.

3.4 L-Systems as Keytool for e-Science

Because of my previous research on expert and lay human cognition, AI, decision-making processes, computational trends in contemporary science and biotechnology dynamics, for several years I have had a strong interest in the interconnections between emotion and reason, exemplified perfectly by contemporary electronic art. But it was after my discussions with Mexican electronic artist Florence Gouvrit about her work *In Silico* [79], that I started to design a solid coherent model of integrative learning that included several kinds of cognitive values (epistemic, aesthetic and emotional).

With that purpose I chose L-systems as a very good way to integrate undergraduate and graduate learning (there are several levels of expertise and difficulty in L-systems use). It is a good way to integrate and teach science in the classroom [73], with recent technological tools [80]. We try to improve the process of students' inquiry with computer-supported learning [81]–[83].

3.4.1 Introducing L-systems

L-systems were introduced in 1968 by Aristid Lindenmayer, a Hungarian theoretical biologists and botanist, as a theoretical framework for studying the development of simple multicellular organisms and were subsequently applied to investigate higher plants and plant organs. After the incorporation of geometric features, plant models expressed using L-systems became detailed enough to allow the use of computer graphics for realistic visualization of plant structures and developmental processes [84]. Not strictly Alife, L-Systems can create realistic looking plant structures using simple recursive rules.

They are a valuable tool for educational purposes because they integrate computer programming, mathematics (fractals theory, algorithms, geometry), Alife and Biology. L-systems enable us to work with not only the knowledge processes mentioned above, but also procedural (quantification and modeling of nature, interdisciplinary approach) and cognitive ones (emotional aspects of L-systems visualization).

There are several geometric features in plant development: (a) *bilateral symmetry* of leaves, rotational symmetry of flowers or helical arrangements of scales in pine cones; (b) with relatively simple developmental algorithms we can describe the fascinating *self-similarity* of plant development.

But, how do L-systems work? Conceived initially as mathematical models of plant development, L-systems have as a central concept that of rewriting. Rewriting is a technique for defining complex objects by successively replacing parts of a simple initial object using a set of rewriting rules or productions. Look for example at Figure 3.2:

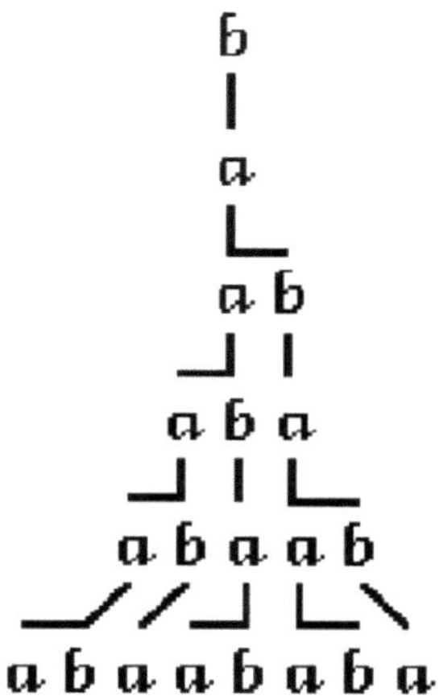

Fig. 3.2. Rewriting rule and derivation

All strings are built of two letters, *a*and *b*. For each letter a rewriting rule is specified: (1) rule *a ? ab* means that we must replace letter *a* by the string *ab*; (2) rule *b?a* means that the letter *b* is replaced by *a*. If we start with *b*, as in the Figure 1, we obtain in 5 steps the string *abaababa*, and so on. The rewriting process starts from a distinguished string called the *axiom*. If we assign Cartesian coordinates and angle increments to the general position of the axiom and subsequent strings, we obtain forms similar to plants.

If we create a fractal in a similar rewriting technique, we obtain the classical snowflake curve of Figure 3.3:

L-systems can be extended to three dimensions representing orientation in space, and can also be randomized stochastically, and can be context-sensitive (that is, the development of a production depends on the predecessor's context). By using all these concepts, L-systems can simulate real plants interacting with each other, simulating various elements of a growing structure as well as the interactions between that structure and the environment. At

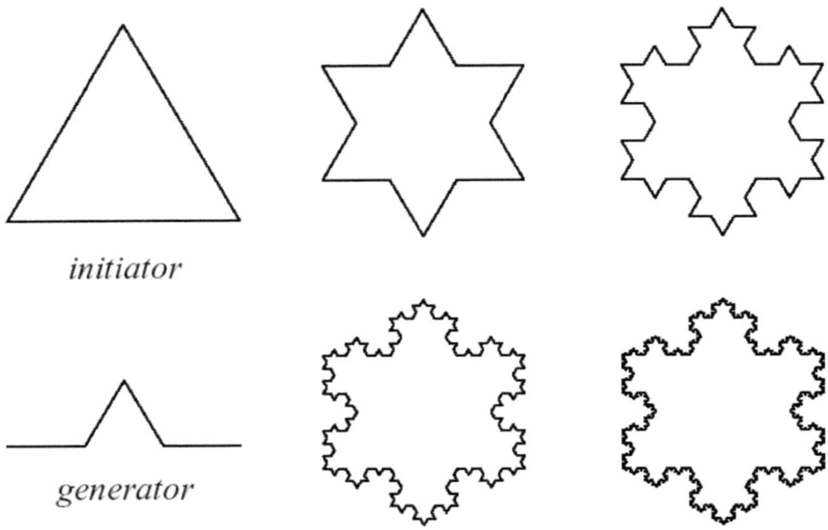

Fig. 3.3. Rewriting rule as fractal

a more advanced level modeling techniques for leaves and petals can also be learned, as in Figure 3.4

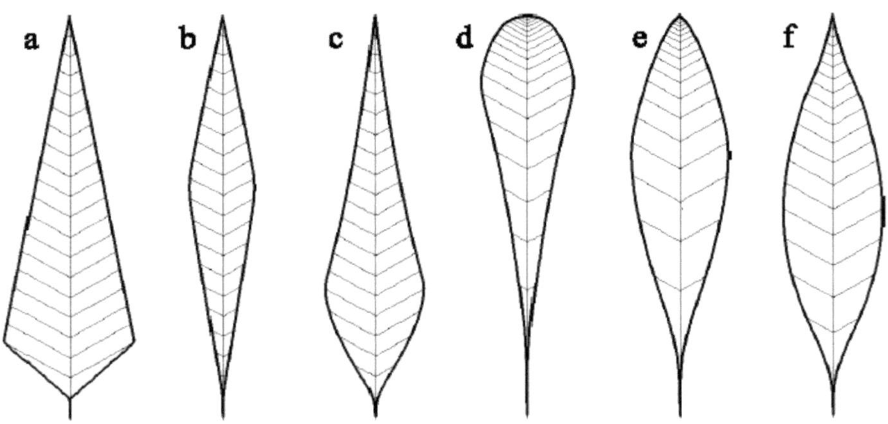

Fig. 3.4. Leaves

L-systems are actually also known as *parametric L-systems*, defined as a set (**G**), and have several components that can be described as:

G $= \{V, S, \omega, P\}$

Where:

V (the alphabet) is a set of symbols containing elements that can be replaced (variables).

$S$is a set of symbols containing elements (constraints) that remain fixed.

$\omega$is a string of symbols from Vdefining the initial state of the system. So, they act as start, axiom or initiator.

$P$is a set of rules or productions defining the way in which variables can be replaced with combinations of constants and other variables. A production consists of two strings: the predecessor and the successor.

To generate graphical images, L-systems require that the symbols in the model refer to elements of a drawing on a computer screen. To achieve that purpose, they use turtle geometry [85]–[87]. Turtle programs provide a graphical interpretation of L-systems, which are special grammars with specific kinds of production rules [88]. The use of Papert & Solomon's turtle in the 1970's for children's computational uses is the best example of a good precedent [89].

Every program uses similar but not identical symbols. In this chapter I propose the use of LSE software because it is very simple and has a very small size (94kb). Nevertheless, there are several programs about L-systems. The computer language *Logo* is best known as the language that introduced the *turtle* as a tool for computer graphics. The crucial thing about the turtle, which distinguishes it from other metaphors for computer graphics, is that the turtle is pointing in a particular direction and can only move in that direction. (It can move forward or back, like a car with reverse gear, but not sideways.) In order to draw in any other direction, the turtle must *first* turn so that it is facing in the new direction.

With this kind of virtual biology software, we can have a virtual laboratory, useful for computerized experimentation, with interactive manipulation of objects, under controlled conditions. We are making e-Science at a very simple level, but it is interdisciplinary, computerized science.

3.4.2 A cheap and easy way to use L-systems

One of the most interesting aspects of L-systems are the common ideas of the scientific and programming communities who have developed these languages: people who believe in open access and freeware. So, nearly all materials we need for our classes (tutorials, software, and website support) are open freely to everyone who wishes to use them.

We can find several papers about this topic and the fundamental and extremely beautiful book (in PDF format), *The Algorithmic Beauty of Plants* at http://algorithmicbotany.org/papers/. All the materials are free and can be easily downloaded. This an important aspect of the present activity: the *open access* culture, in which we can find important concepts such as 'free software', 'copyleft', 'GPL' (*General Public License*),..., If the teacher explains the origin of all the materials used, the student can understand the challenging nature of scientific research inside an e-Science paradigm, where the information is experienced, not possessed. This is also a good opportunity to talk with students about the nature of scientific research, and the radical transformations produced in it by the Internet. Beyond downloading of music, video and other materials through P2P, the Net enables a different and collaborative way to create knowledge. Open source platforms like Linux are an example of distributed, coordinated, virtual and social construction of knowledge. There are several software programs to develop L-systems as listed in Table 3.1 and Tabletab2:

Table 3.1. Several L-systems programs and their platforms

Program	Creator	Platform
Fractint	Stone Soap Group	Windows Mac Linux
The Virtual Laboratory	The Biological Modeling and Visualization Group	Windows
LS Sketchbook	Roberto S. Ferrero	Windows
LSE	James Matthews	Windows
L-systems 4.0.1.	T. Perz	Windows
Lmuse	D. Sharp	Windows
Lyndyhop	-	Windows Mac

Table 3.2. Categories of L-systems software and where to find it

Program	Software	Website
Fractint	Freeware	spanky.triumf.ca/www/fractint/fractint.html
Virtual Laboratory	Evaluation Ver.	algorithmicbotany.org/virtual_laboratory/
LS Sketchbook	Freeware	coco.ccu.uniovi.es/malva/sketchbook/
LSE	Freeware	www.generation5.org/content/2002/lse.asp
L-systems 4.0.1.	Freeware	www.geocities.com/tperz/L4Home.htm
Lmuse	Freeware	www.geocities.com/Athens/Academy/8764/lmuse/lmuse.html
Lyndyhop	Freeware	www.lab4web.com/chelmiger/lyndyhop/

I recommend two of them, LS Sketchbook and LSE. Both have a good interface and are compatible with several PC operating systems. Zipped, LS Sketchbook has a size of 2387kb, and LSE 94kb. If you have memory problems, the best option is therefore, LSE.

Perhaps for the purpose of this paper, it would be better start with LSE and use LS Sketchbook just for more advanced levels.

3.4.3 How to use LSE: easy programming

LSE, created by the prolific James Matthews, allows for up to 26 different rules, loading and saving of rules, zooming, panning and depth adjustment, recursive drawing instead of preprocessing the string and an additional L-Systems support.

According to its creator [90], LSE is simple to use: you can try loading in some of the preset systems (*.lse) and then playing about with the parameters. It is an intuitive approximation to LSE. All parameters can be accessing using the "Edit, L-Systems..." command. This will bring up a dialog box with all the parameters (angle, strings, initial angle, segment and step size), the axiom and rules. Once the system has been drawn you can pan the system about by dragging with your mouse. Note that holding down "Ctrl" while dragging will force the L-system to start drawing from the mouse position. For those of you with an IntelliMouse, you can increase the segment size by scrolling up and down, if you hold down the "Ctrl" key and scroll it will increase the depth. Note this has to be turned on in the options, since it is easy to draw at a depth that takes too long to execute, effectively hanging LSE. The interface of LSE, as Figure 3.5:

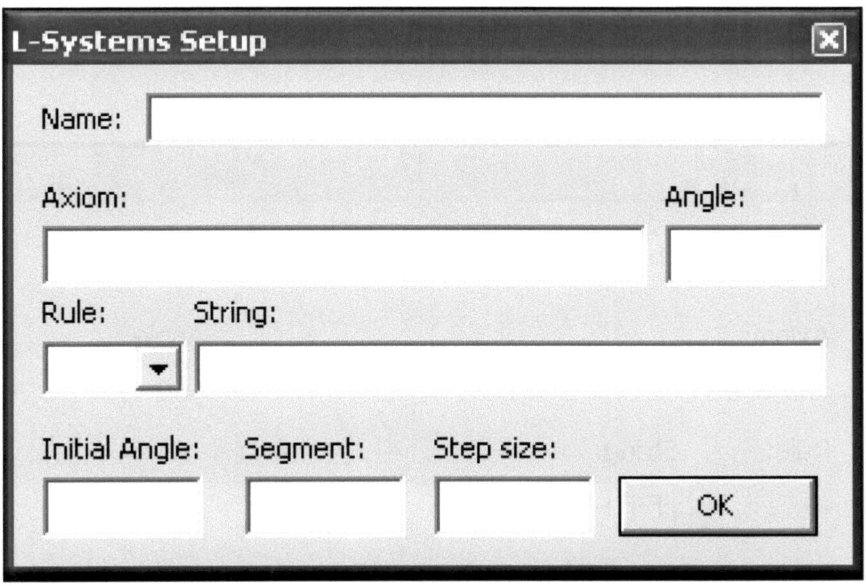

Fig. 3.5. LSE setup programming interface

L-Systems Explorer allows you to assign up to 26 recursive rules (A, B, C...Z). When the rules are expanded, the following commands are understood:

F: Move forward by 'x' amount, drawing a line.

G : Move forward by 'x' amount, without drawing line (note, f can be used too)

+: Rotate by theta degrees in the positive direction.

- : Rotate by theta degrees in the negative direction.

[: Push the draw state (position and angle) on the stack.

]: Pop the draw state (position and angle) off the stack.

|: Move forward by stepped amount.

The user specifies 'x' and theta. You may also use numbers to specify multiple angle increments or decrements - therefore, the following two statements are equivalent: "4+F" and "++++F".

A final clarification, the '|' command is a little strange. It differs from the F command because *it is not recursively expanded.* When an F is reached, it is often substituted for whatever the F-rule specifies, and the segment length is decreased by a certain percentage, therefore the only F-commands that get interpreted are the commands at the final depth. This is not so with the '|' command, allowing a little more control over the segment lengths.

3.4.4 Easy results and advanced possibilities

LSE is simple L-systems software, but useful to start to learn. For example, we can produce the snowflake fractal of Figure 2, in the way showed by Figure 3.6:

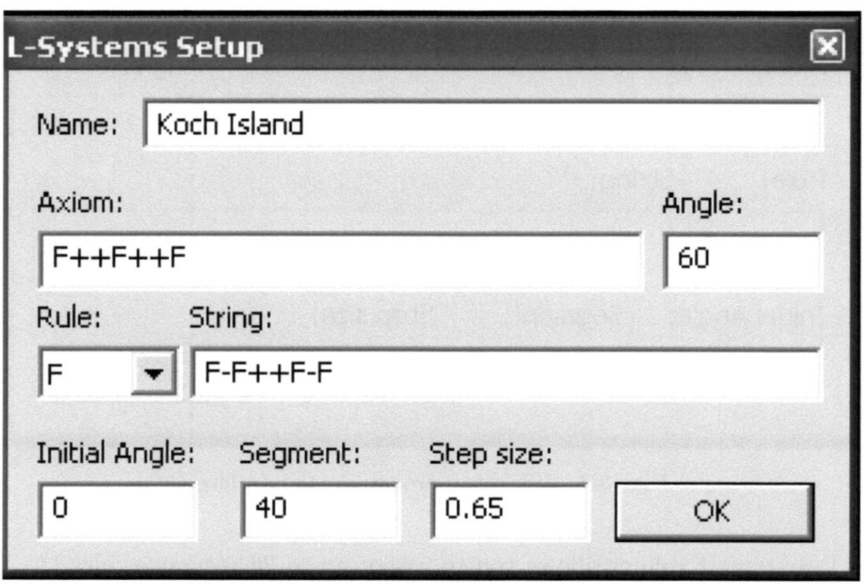

Fig. 3.6. Programming the Snowflake with LSE

I can show you a typical plant form (Figures 3.7 and 3.8) that I created easily by making modifications of the bush-ex1.lse file included with the LSE creator, in just 5 minutes of trial and error activity. The first figure is the obtained vegetal image, whereas second it is a sample of the instructions necessary to create it:

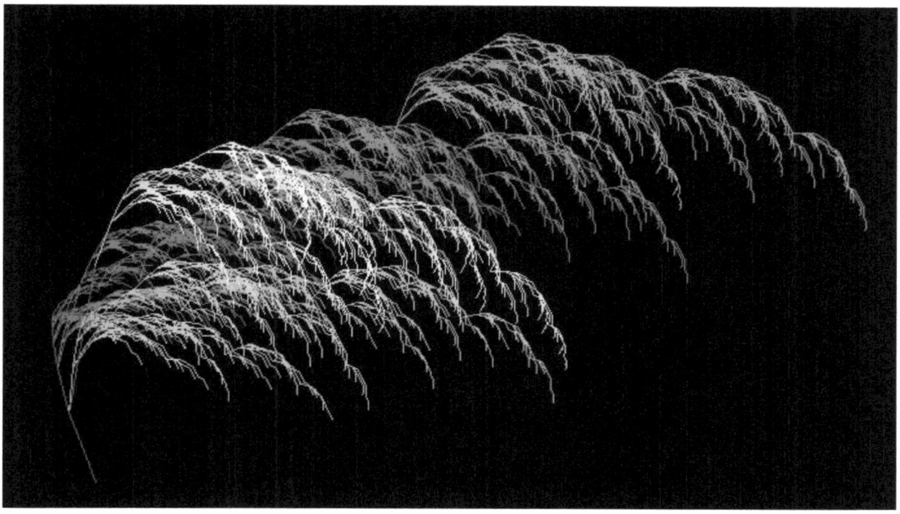

Fig. 3.7. Example of my own static Alife creation with LSE

So, the question is that with LSE or similar software, students can do biology, mathematics, computer programming and art at the same time. We have seen in previous analysis that a computational way to obtain knowledge exists, with the presence of emotional-artistic values. That is the real e-Science.

LSE constitutes a *static* first step to achieve a more dynamical model of biological systems. But, in the end, it is Alife. Perhaps: 'Alife for beginners'. At a more advanced level, there are more evolved programs like LS Sketchbook which enable us to develop more sophisticated L-systems or Java software based on cellular automata [91] to reproduce living systems.

3.4.5 What are the objectives of creating images similar to the previous one?

Scientists often create visual images of what they cannot see or adequately comprehend: from molecules and nanostructures to cosmic reality; of phenomena both real and abstract, simple and complex [92].Often in a parallel manner, science educators use images created by scientists or virtual images they fashion themselves, to extend the intellectual horizons of their students. But

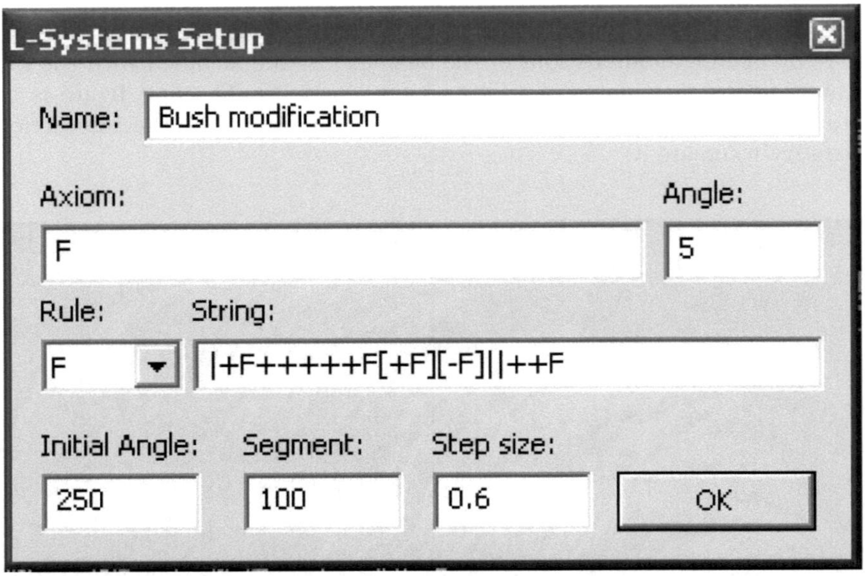

Fig. 3.8. Screen capture with the programming rules necessary to obtain the Figure 3.7

it is also possible to allow the students to create their own images, as happens with LSE and Alife systems. Interactive, computer based animations and visualizations have equipped students and teachers to see and understand complex science concepts. And that kind of learning interacts with holistic development [http://community.middlebury.edu/~grc/].

I recommend applying L-systems to the classroom in this sequence:

1st Analyze geometrical aspects of plants: leaves, photographs, graphics...

2nd Introduce basic ideas about fractals and mathematical models of life.

3rd You can use the example of Artificial Intelligence and geometric basis present in computer games, something very familiar to them.

4th Teach the students the fundamentals of L-systems.

Use, in groups, of LSE: first of all make the whole process:
localize the program, download freely, install it and load in some
of the preset systems (*.lse) and then play about with the parameters.

5th Share all the group results in a common viewing. Comment on why they are different and the underlying geometrical reasons for these differences.

6th Discuss the value of a scientific model, and if differences exist between experimental and virtual models. At this point all the group is thinking about epistemology and scientific methodology. It would be a good idea for the students to see examples of contemporary projects of computerized scientific simulations (climate, chemistry, genome...).

7^{th}It would also be very interesting for the teachers to show their students websites were they can also look at advanced L-systems, created by professionals or enthusiasts of these languages.

8^{th}To be able to use the Network strategies of contemporary Science I recommend the creation of a website of LSE creations made by students, to allow the exchange opinions, ideas and results with other schools. It could be the start of an integrative, open, and collaborative way to learn and teach science. The idea is to produce science by making it.

9^{th}If the group can master LSE, you can try to change to LS Sketchbook or search for other Alife software, which can enable the creation of dynamical Alife.

3.4.6 Learning by doing

One of the questions that could be asked by the reader is "well, we know how to use the LSE software, and have learned several ideas about the epistemological validity of simulations... but are our students really learning anything (and what) with this process?". My answer is: 'yes'. For philosophers of science it is clear that scientific activity includes both open rules and strategies as well as tacit knowledge [93]–[94]. Cognitive abilities (related to practices) are as important as mental skills (coordinated by research strategies). All these domains of human activity can be developed creating islands of expertise with LSE (and other Alife systems) software.

Crowley and Jacobs [95]:333, define an island of expertise as "any topic in which children happen to become interested and in which they develop relatively deep and rich knowledge." These areas of connected interest and understanding, they suggest, create "abstract and general themes" that become the basis of further learning, both within and around the original topic area, and potentially in domains further afield. Starting from the geometrical nature of plants, and with group work strategies, these islands of expertise emerge creating at the same time 'epistemic frames'. According to Shaffer [96]: 227: "epistemic frames are the ways of knowing associated with particular communities of practice. These frames have a basis in content knowledge, interest, identity, and associated practices, but epistemic frames are more than merely collections of facts, interests, affiliations, and activities".

Working with Alife software like LSE, students create images with rich meanings: that is, the meanings of images involve relations to other things and ideas. These relations are symbolic in the sense that they are matters of convention that involve some degree of arbitrariness. Some conventions exploit natural correspondences that facilitate image understanding [97]. For example: the *in silico* development of plants, the geometrical nature of plants, the joint action of computers and scientific research ("behind every great human is a great computer/software"), basic ideas of programming, the culture of free and open access software, the virtues of visual thinking and philosophical aspects of virtual science,...

Creating an image with LSE, the students not only put together academic knowledge (from biology, informatics or mathematics fields) but also engage in an activity of active integration of ideas and practices that design an interdisciplinary attitude on scientific research. A good visual model can also help teams realize the synergy of their collective knowledge (working in teams or sharing collectively the obtained knowledge), and inspire them to move forward with a shared sense of understanding, ownership and purpose while at the same time they think about the scope of models. And around the Alife simulation by LSE, they create an island of expertise which enables them to integrate ideas from different fields at a similar practical level. It's easier to make an alife simulation first than to analyze all the theoretical background that constitute them. The interest of the visualization and the rapid changes produced with LSE, allow posterior discovery of their deeper roots.

And we can ask ourselves: what kind of scientific interpretation the creation of "alife" structures can authorize? The answer is simple: true life, if the model is a good model. In our case, LSE provides a limited modelization of a living plant. But that is not the real point. Although, they have learned plenty of things working with LSE, the teacher and the learners have also discussed important questions such as: have they really understood the epistemological value of a scientific model? Are there several levels of veracity and similitude between models and reality? Are real models different from virtual ones? How can we decide about the truthfulness of a model? And the answer is an open exercise of critical thinking, guided by a pragmatic principle: "does it work?"

Moreover this is not a teacher-centered activity, but a pedagogical model in which the teacher acts as a catalyzer of the students' cognitive capacities through an imaging software (a true extension of the students' mind). Teachers create a rich learning space in which learners develop an active role by creating their own knowledge. And that is *critical knowledge*, because both teacher and students have discussed the meaning of the models and values implied in them. From a socially-situated conception of learning, toward viewing intelligence as a distributed achievement rather than as a property of individual minds, this is a dialogical process of knowledge construction. That activity, creates dynamic knowledge, because activity is enabled by intelligence, but that intelligence is distributed across people (teacher and students), environments (class, books, Internet and software), and situations (discussions, explanations, training, uses), rather than being viewed as a resident possession of the individual embodied mind [98].

Finally, we can ask ourselves how we can be objectively sure that by using these same devices, students will in fact integrate different types of knowledge? We can look at several indicators:

- Can students use the program effectively and explain how they do it?
- Are they able to explain how the plants grow and why they manifest a special geometrical structure?
- Are they conscious about the nature of contemporary science?

- Do they understand the epistemological value of a model, and a virtual model?
- Has their interest in the different blended fields implied in the Alife model (biology, mathematics, informatics) increased ?.
- Are they interested in developing more projects about Alife (including analysis of videogames)?
- Do they – or a substantial number of them - obtain better scores in disciplines that have been practiced?

All these questions are related to cognitive abilities to reach knowledge: rules, habits, empathy, emotion, interest, ... which produce not only meaning but also an intentional behaviour oriented to achieve more meaning through learning-by-doing.

One way to respond to these questions is to organize a final group discussion. At the same time we can design a simple survey. And, finally, look at the posterior curricula evolution of students involved in the activity (by means of the supervision of their respective teachers).

3.5 Summary

We have seen that Alife worlds (static or dynamic) are very useful in the process of developing an integrative science. They fit well with contemporary trends in scientific enterprise (interdisciplinary, highly computerized, Network strategies) and with the latest cognitive models (which consider the crucial role of emotional or non-epistemic values). Electronic art applied to Science Education, is not an external work made by the artist but requires the active involvement of the public: making beautiful L-systems, our students can learn at the same time mathematics, basic programming, biology, the new e-Science and art. Working with Alife systems, the emotional and cognitive necessities of our students are brought together in an intuitive and fun tool. The aesthetics of the results comprise user-friendly knowledge and emotion. So, it is a better way to learn Science.

Acknowledgements

I would like to thank: Florence Gouvrit for insightful comments and her stimulating electronic art, Mercè Izquierdo for her ever interesting comments about Science Education, Roberto S. Ferrero for her suggestions, James Matthews for his beautiful free software, my "Philosophy and Computing" students for their ideas and suggestions and finally, UAB's Philosophy Department for allowing me to know a new generation of young students every year. Finally, I thank the anonymous reviewer for her/his truly helpful comments and criticism.

This research has been developed under the main activities of the TECNOCOG research group (UAB) about Cognition and Technological Environments, [HUM2005-01552], funded by MEC (Spain).

References

1. J.D. Watson, and F.H.C. Crick, "Molecular Structure of Nucleic Acids. A Structure for Desoxyribose Nucleic Acid", *Nature*, Vol. 4356, April 25^{th}1953, pp. 737-738.
2. J.C. Venter, J.C. *et al*, "The Sequence of the Human Genome", *Science*, Vol. 291, 2001, pp. 1304-1351.
3. National Research Council, *Bio2010: Transforming Undergraduate Education for Future Research Biologists*, National Academy Press, Washington, D.C., 2003.
4. J.L. Gross, R. Brent, and R. Hoy, "The Interface of Mathematics and Biology", *Cell Biology Education*, The American Society for Cell Biology, Vol.3, pp. 85-92, 2004.
5. W. Bialek, and D. Botstein, "Introductory science and mathematics education for 21st-century biologists", *Science*, Vol. 303, February 6th 2004, pp. 788-790.
6. http://binfo.ym.edu.tw/edu/seminars/seminar-041002.pdf. See also footnote number [12] pp. 7.
7. R. Logan, *The sixth language: Learning a living in the internet age,* The Blackburn Press, New Jersey, 2000.
8. R.Greenberg, J. Raphael, J.L.Keller and S. Tobias, "Teaching High School Science Using Image Processing: A Case Study of Implementation of Computer Technology", *Journal of Research in Science Teaching*, vol 35, no. 3, 1998, pp. 298-327.
9. American Association for the Advancement of Science,*Benchmarks of science Literacy,* Oxford University Press, NY, 1993.
10. NRC, *National science education standards,* National Academy, Washington, D.C., 1996.
11. P.B. Hounshell, S.R.Hill, "The microcomputer and achievement and attitudes in high school biology", *Journal of Research in Science Teaching*, vol.26, 1989, pp.543–549.
12. Z. Zacharia, "Beliefs, Attitudes, and Intentions of Science Teachers Regarding the Educational Use of Computer Simulations and Inquiry-Based Experiments in Physics", *Journal od Research in Science Teaching*, vol. 40, no. 8, 2003, pp. 792-823.
13. M. Taylor, and P. Hutchings, *Integrative Learning: Mapping the Terrain*, Association of American Colleges and Universities and The Carnegie Foundation for the Advancement of Teaching, Washington, D.C., 2004. [Available at http://www.carnegiefoundation.org/IntegrativeLearning/mapping-terrain.pdf].
14. G. Fauconnier, M.Turner, "Conceptual integration networks", *Cognitive Science*, Vol. 22, no. 2, 1998, pp.133–187.
15. P. Bell, E.A. Davis, & M.C. Linn, "The Knowledge Integration Environment Theory and Design", in *Proceedings of the Computer Supported Collaborative Learning Conference (CSCL '95: Bloomington, IN).* Mahwah, NJ: Lawrence Erlbaum Associates, 1995, pp. 14-21.

16. M. Gibbons, C. Limoges, H. Nowotny, S. Schartzman, P. Scott, and M. Trow, *The new production of knowledge: The dynamics of science and research in contemporary societies,* Sage Publications, London, 1994.
17. M. Castells, The Rise of the Network Society, Blackwell, Oxford, 1996.
18. Most references to e-Science come from the inedited paper presented at iC&P Congress by J. Vallverdú, "Computational epistemology and e-Science", held in Laval, France, May 2006.
19. A. Clark, and D. Chalmers, "Embodied, situated and distributed cognition", in W. Bechtel & G. Graham (eds.) *A Companion to Cognitive Science*, Backwell, Malden (MA), 1998.
20. P. Humphreys, Extending*Ourselves. Computational Science, Empiricism and Scientific Method*, Oxford University Press, USA, 2004.
21. E. Hutchins, *Cognition in the Wild.* MIT Press, Cambridge (MA), 1995.
22. D.A. Norman, *The Invisible Computer*, MIT Press, Cambridge (MA), 1997.
23. L. Floridi, "Information Ethics: An Environmental Approach to the Digital Divide", *Philosophy in the Contemporary World*, Vol. 9 (1: Spring-Summer), 2001.
24. L. Floridi,"What is the philosophy of information?", *Metaphilosophy*, Vol.33(1-2), 2002, pp. 123-145.
25. L. Floridi, Luciano (2003), "Two approaches to the philosophy of information", *Minds and Machines*, Vol. 13, No. 4, 2003, pp. 459-69.
26. T.W. Bynum, and J.H. Moor, *The Digital Phoenix. How Computers Are Changing Philosophy*, Blackwell Publishers, UK, 1998.
27. P. Thagard, *Computational Philosophy of Science*, MIT Press, Cambridge (MA), 1988.
28. G.P. Landow, (ed.), *Hyper/Text/Theory*, Johns Hopkins University Press, USA,1994.
29. J-F. Rouet, *Hypertext and Cognition*, Lawrence Erlbaum Associates, Inc, USA, 1996.
30. J. Vallverdú, "Hypertextual Thoughts", *Revista Portuguesa de Filosofia, In Press*, 2005.
31. J.D. Bransford, A.L. Brown, R.R. Cocking (eds.), *How people learn: Brain, mind, experience, and school*, National Research Council/National Academy Press: Washington, D.C., 1999.
32. D. F. Halpern, M. K. Hakel, "Applying the science of learning to the university and beyond: Teaching for long-term retention and transfer", *Change*, Vol. 35, No. 4, 2003, pp.21-29.
33. R. Glaser, "Education and thinking: The role of Knowledge", *American Psychologist*, vol. 39, pp. 93-104, 1984.
34. D.J. Hacker (ed.) *Metacognition in Educational Theory and Practice* Lawrence Erlbaum Associates, London, 1998.
35. L.B. Resnick, "Mathematics and science learning: A new conception", *Science*, vol. 220, 1983, pp. 477 - 478.
36. J.R. Carbonell, *Mixed-initiative man-computer instructional dialogues.* Doctoral Dissertation, Massachusetts Institute of Technology, Cambridge, Massachusetts, 1970.
37. R.A. Khuwaja, *A model of tutoring facilitating knowledge integration using multiple models of the domain,* Doctoral Dissertation, Illinois Institute of Technology, Chicago, 1994.

38. D. A. Schon, *The reflexive practitioner: How professionals think in action*, Basic Books, New York, 1983.
39. A. Clark, D.J. Chalmers, "The Extended Mind", *Analysis*, Vol. 58, No. 1, pp. 7-19, 1998.
40. P. McClean *et al*, "Molecular and Cellular Biology Animations: Development and Impact on Student Learning", *Cell Biology Education*, The American Society for Cell Biology, Vol. 4, Summer 2005, pp. 169-179.
41. K.W. Brodie *et al*, *Scientific Visualization*, Springer-Verlag, Berlin, 1992.
42. D.N. Gordin, and R.D. Pea, "Prospects for scientific visualization as an educational technology", *J.Learn. Sci.*, Vol. 4, 1995, pp. 249-279.
43. R. Damasio, *Descartes' Error: Emotion, Reason, and the Human Brain*, Harper, London, 1994.
44. P. Thagard, *Hot Thought: Mechanisms and Applications of Emotional Reason*, MIT Press, Cambridge (MA), in press. Nevertheless, Dr. Thagard has been making research about hot cognitive values in scientific practices from 1992.
45. D.A. Norman, *Emotional design. Why we love (or hate) everyday things.* Basic Books, USA, 2004.
46. N. Sinclair, "The Roles of the Aesthetic in Mathematical Inquiry", *Mathematical Thinking and Learning*, Lawrence Erlbaum Associates, Vol. 6, No. 3, 2004,pp. 261- 284.
47. M. Kemp, *Visualizations. The Nature Book of Art and Science*, Oxford: OUP, 2000.
48. M. Kemp, "From science in art to the art of science", *Nature*, Vol. 434, March 17th 2005, pp. 308-309.
49. M. Claessens (ed.), "Art & Science", *RTDinfo Magazine for European Research* (Special Edition), European Commission, March 2004, pp. 1-44. [Available at www.europa.eu.int/comm/research].
50. J.H. Mathewson, "Visual-Spatial Thinking: An Aspect of Science Overlooked by Educators", *Science Education,* Vol. 83, 1999, pp. 33-54.
51. P.M. Churchland, *The engine of reason, the seat of the soul*, MIT Press, Cambridge (MA), 1995.
52. C. Comoldi, and M.A. McDaniel (eds.), *Imagery and cognition*, Springer, New York, 1991.
53. P.J. Hampson, D.F. Marks, and J.T. Richardson (eds.) *Imagery: Current developments*, Routledge, London, 1990.
54. S.M. Kosslyn, *Image and brain. The resolution of the imagery debate*, Fress Press, New York, 1994.
55. D. Marr, *Vision*, W.H. Freeman, New York, 1982.
56. J. Piaget, and B. Inhelder, *Mental imagery in the child: A Study of the development of imaginal representations*, Routledge & Kegan Paul, UK, 1971.
57. S. Pinker, *How the mind works*, Norton, New York, 1997.
58. S. Ullman, *High-level vision*, MIT Press, Cambridge (MA), 1996.
59. B. Tversky, J.B. Morrison, M. Betrancourt, "Animation: Can it facilitate?", *International Journal of Human-Computer Studies*, vol. 57, 2002, pp.247-262.
60. J.K. Gilbert (ed.), Visualization in Science Education, Series: Models and Modeling in Science Education , Vol. 1, UK, Springer Verlag, 2005.
61. D.N. Gordin & R.D. Pea, "Prospects for scientific visualization as an educational technology", *Journal of the Learning Sciences*, vol. 4, 1995, pp.249-279.
62. http://www.idiagram.com/ideas/knowledge_integration.html. Accessed on May, 26th 2006.

63. C.G. Langton (ed.), *Artificial Life*, Redwood City, Addison-Wesley, 1989.
64. R. Brooks, "The relationship between matter and life", *Nature*, Vol. 409, January 18th 2001, pp. 409-411.
65. C. Adami, *Introduction to Artificial Life*, Springer Verlag, NY, 1998.
66. J.L. Casti, "The melting-pot that is Alife", *Nature*, Vol. 409, January 4th 2001,pp.17-18.
67. http://www.alife.org/.
68. S.A. Benner, "Act Natural", *Nature*, Vol. 421, January 9th 2003, pp. 118.
69. M.A. Bedau, "Philosophical Aspects of Artificial Life", in F.J. Varela, and P. Bourgine (eds.), *Toward a Practice of Autonomous Systems: Proceedings of the First European Conference on Artificial Life*, MIT Press, Cambridge (MA), 1992, pp. 494-503.
70. M. A. Boden (ed.), *The Philosophy of Artificial Life*, Oxford University Press, Oxford, 1996.
71. D.C. Dennet, "Artificial Life as Philosophy", *Artificial Life*, Vol. 1, No. 3, 1994, pp. 291-292.
72. H.H. Pattee, "Artificial life needs a real epistemology", in F. Moran, A. Moreno, J.J. Morelo, P. Chacon (eds.), *Advances in Artificial Life*, Springer Verlag, Berlin, pp. 23-28.
73. H. Putnam, "Robots: Machines or artificially created life?", *Journal of Philosophy,* Vol. LXI, No. 21, November 12th 1964, pp. 688-691.
74. G.F. Miller, "Artificial life as theoretical biology: how to do real science with computer simulation", *Cognitive Science Research Paper no. 378*, School of Cognitive and computing Sciences, University of Sussex, Brighton, UK, 1995.
75. M. Whitelaw, *Metacreation: Art and Artificial Life*, MIT Press, Cambridge (MA), 2004.
76. L. Candy, and E. Edmonds, *Explorations in Art and Technology*, Springer Verlag, UK, 2002.
77. M. Boden, *Dimensions of Creativity*, MIT Press, Cambridge (MA), 1994.
78. F. Varenne, "What does a computer simulation prove ?, *Simulation in Industry*, Proc. of The 13th European Simulation Symposium, Marseille, France, October 18-20th , 2001, ed. by N. Giambiasi and C. Frydamn, SCS Europe Bvba, Ghent, 2001, pp. 549-554.
79. www.gouvrit.org. I thank Florence for her time and the opportunity to discuss with her several ideas about electronic art, technology and science, developed by the artist and Liliana Quintero, both researchers from Centro Multimedia, Centro Nacional de las Artes, México D.F. Her work *in Silico* has been very stimulating for me.
80. M.S. Donovan, and J.D. Bransford (eds.), *How Students Learn: Science in the Classroom*, NRC/NAS, Washington, D.C., 2005.
81. E.M. Coppola, *Powering Up: Learning to Teach Well with Technology*, Teachers College Press, USA, 2004.
82. K. Hakkarainen, and M. Sintonen, "The Interrogative Model of Inquiry and Computer-Supported Collaborative Learning", *Science & Education*, Vol. 11, 2002, pp. 25-43.
83. M. Linn, and S. His, *Computers, Teachers, Peers: Science Learning Partners*,Lawrence Erlbaum, USA, 2000.
84. W. McKinney, "The Educational Use of Computer Based Science Simulations: some Lessons from the Philosophy of Science", *Science & Education*, Vol. 6, 1997,pp. 591-603.

85. P. Prusinkiewicz, and A. Lindenmayer, *The Algorithmic Beauty of Plants*, Springer-Verlag, New York, 1990. [Available as a free PDF file at http://algorithmicbotany.org/papers/. Most illustrations of the present chapter, if it is not otherwise indicated, belong to this excellent work].
86. H. Abelson, and A.A. diSessa, *Turtle Geometry. The Computer as a Medium for Exploring Mathematics*, MIT Press, Cambridge (MA), 1981.
87. B. Harvey, "Turtle Geometry", *Computer Science Logo Style,* Vol. 1: *Symbolic Computing* 2/e, MIT Pres, Cambridge (MA), 1997. [Available at http://www.cs.berkeley.edu/~bh/pdf/v1ch10.pdf]
88. P. Prusinkiewickz, "Graphical applications of L-systems", *Proceedings of Graphical Interface 86 and Vision Interface 86*, 1986, pp. 247-253.
89. S. Papert, and C. Solomon, "Twenty things to do with your computer", *Educational Technology*, Vol. 9, No. 4, 1972, pp.39-42.
90. http://www.generation5.org/content/2002/lse.asp, all the LSE technical instructions have been extracted from the quoted website.
91. S. Wolfram, *A New Kind Of Science*, Wolfram Media, Inc., USA, 2002.
92. R.M. Friedhoff, *Visualization : the second computer revolution*, W.H. Freeman and company, NY, 1991.
93. M. Polanyi, *The Tacit Dimension*, Anchor Books, NY, 1967.
94. H.M. Collins, "The TEA Set: Tacit knowledge and scientific Networks",*Science Studies*, vol. 4, 1975, pp.165-185.
95. K. Crowley, M. Jacobs, "Islands of expertise and the development of family scientific literacy", in G.Leinhardt, K. Crowley, & K. Knutson (Eds.), *Learning conversations in museums*, Lawrence Erlbaum, Mahwah (NJ), 2002.
96. D.W. Shaffer, "Epistemic frames for epistemic games", *Computers & Education*, vol. 46, 2006, pp. 223–234.
97. D.J. Waddington, "Molecular Visualization in Science Education", Report from the Molecular visualization in science education workshop, NCSA Access Center, Arlington, (Sponsored by the National Science Foundation) VA January 12-14, 2001.
98. D.R. Pea, "Augmenting the discourse of learning with computer-based learning environments", in E. De Corte, M. Linn, & L. Verschaffel (Eds.), *Computer-based learning environments and problem-solving* (NATO Series, subseries F: Computer and System Sciences). New York: Springer-Verlag GmbH, pp. 313-343, 1992.

4

Pedagogic Strategies Based on the Student Cognitive Model Using the Constructivist Approach

Louise Jeanty de Seixas[1], Rosa Maria Vicari[2], and Lea da Cruz Fagundes[3]

[1] Post-graduation Program (Computer in Education), Federal University of Rio Grande do Sul – UFRGS, Porto Alegre-RS, BRAZIL louise.seixas@ufrgs.br
[2] Informatics Institute, Federal University of Rio Grande do Sul – UFRGS, POBox:15064, 91501-970,Porto Alegre, RS, Brazil rosa@inf.ufrgs.br
[3] Post-graduation Program (Computer in Education), Federal University of Rio Grande do Sul – UFRGS, Porto Alegre-RS, BRAZIL leafagun@vortex.ufrgs.br

This study is intended to assess whether it is possible to design pedagogic strategies based on models of conscience awareness and use them, by means of intelligent agents. Piaget's theory of knowledge construction was our theoretical basis, specially the equilibration theory, and the studies about cognitive conducts. AMPLIA, an intelligent learning environment which uses bayesian networks for the knowledge representation of the student and the expertise was used for this study: Their probabilistic intelligent agents evaluate the student's network, make inferences about the student's actions, and select the pedagogic strategy that is more likely to be useful. These strategies are aimed to make the student conscious about a study case – the student can make hypothesis and is able to test them. The strategies were based on the procedures studied by Piaget, and they are presented to the student as tactics with different autonomy levels. Some examples of strategies are presented, and also the follow up of a complete interaction process of a student, from the lowest to the highest cognitive stage. This dynamic process among the students and AMPLIA is very similar to the interaction among the students and their teachers, so we can conclude that intelligent agents can use pedagogic strategies based on student's cognitive model, as proposed.

4.1 Introduction

This article is part of a Doctoral dissertation - *Pedagogic strategies for an intelligent multi-agent learning environment – AMPLIA* [31] presented in 2005

Louise Jeanty de Seixas et al.: *Pedagogic Strategies Based on the Student Cognitive Model Using the Constructivist Approach*, Studies in Computational Intelligence (SCI) **44**, 77–102 (2007)
www.springerlink.com

at the Post-graduation Program (Computer and Education), at Universidade Federal do Rio Grande do Sul (UFRGS) Brazil - with the hypothesis that it is possible to create pedagogic strategies based on models of consciousness levels and that intelligent agents are able to use such strategies. For that end, our study was based on the genetic epistemology developed by Jean Piaget [24], once such a theory describes and allows one to follow each stage in the knowledge construction process and to identify different cognitive states of the subject, or learner, in the specific case of our study.

Roesler and Hawkings [27] define intelligent agents as independent computing programs that act in software environments as operating systems, databases or networks. Fenton-Kerr [6] approaches the development of pedagogic agents, i.e., intelligent agents developed with educational goals. This author mentions from online help to customized help offers, based on expert systems. The agent's actions will depend basically on the type of relation between student and agent, similarly to the relation between student and professor: (a) Hierarchical, the teacher being the highest authority, aiming at conveying knowledge and making quantitative assessments, or (b) Heterarchical (non-hierarchical), when there is a collaborative work between teacher and students, based on knowledge building and process assessment. Sometimes such agents may be complex programs, and are named Intelligent Tutoring Systems (ITS).

Self [32] says that the pedagogic role of an ITS, from the constructivist point of view, is to provide spaces for interaction to the learner based on some model of the affordances of potential situations. Baylor [1] highlights the functions of an intelligent agent as a cognitive tool: as an information manager, as a pedagogic agent and as an environment manager for the student. The agent, therefore, must follow a student model that can be built from different perspectives. According to Halff [10], the intervention of a pedagogic agent can be through two ways: i) Model tracing, used when the student deviates from the problem solution; and ii) Issue-based tutoring, when the agent identifies opportunities to intervene.

Bull and Pain [3] approach the importance of a student model that counts on the learner's participation, and authors such as Bercht [2] and Jaques [13] develop the research of agents with affective features and emotional states. Mullier and Moore [17] point out that the majority of agents are limited because they use a student model based on domain-specific rules. They suggest that modeling should be performed through neural or probabilistic networks, which are closer to the human behavior. Some examples of this modeling are: Andes [4] in which the student model is built through a Bayesian network (BN)[4]; the work developed by Murray and VanLehn [19] with a dynamic

[4] A Bayesian network is a direct acyclic graph where nodes are random variables and arcs represent direct probabilistic dependence relations among the nodes they connect. The strength of the relationship of Xi with pa(Xi), its parents (nodes with arcs that arrive in Xi), is given by P(Xi | pa(Xi)), the conditional probability distribution of Xi given its parents.

decision network; Capit [16], which uses BNs based on models built by experts and decision theories, to model the student and to guide tutor's actions.

Our proposal is to approach the design of strategies and tactics for a pedagogic agent, based on a student cognitive model that follows the constructivist theory. The student model is inferred by intelligent agents and is probabilistically represented through BNs. We present the theory that supports the construction of the student cognitive model, and we introduce AMPLIA, an intelligent multi-agent environment and describe its intelligent agents – this is where the strategies were implemented. After that we discuss about the pedagogic strategies and variables considered in their selection and present an application of such strategies in AMPLIA and the discussions about probable cognitive models. At the end, there are final considerations, the summary and references.

4.2 Theoretical basis

The pedagogic model of AMPLIA [29] is based on Piaget's theory of knowledge construction [24]. The concept of structure, therefore, is fundamental, as the genetic epistemology presents the process of cognitive structures formation. As it follows, the concepts of assimilation, accommodation and equilibration, as well as the cognitive conducts as described by Piaget, will be the basis upon which we will discuss the consciousness awareness process which, at the end, is what we expect the learner is able to reach.

4.2.1 Cognitive structures

The development of these structures can be observed in very well defined stages since the infant is born. At the first stage - the sensorimotor - actions do not have a meaning themselves (they are not meaningful); this only happens when the actions became coordinated and conscious. The association of coordinated actions leads to more complex and intentional actions upon the object. At this moment, the objects or the thought start to be represented. Such representation takes place through semiotic elements, originating the concept and allowing for internalization.

The following stage typically occurs from ages 2 to adolescence. In this phase, the subject needs the presence of the object, to operate, (to use logical operations or principles), i.e., to use classification, seriation and correspondence, which are closed systems – named cognitive structures. Further, lower structures are used to construct more complex structures. This is the logical stage, which is characterized by the possibility of operating with differentiation, reversibility and transitivity. Notions of causality and function are also developed in this phase through spatial operations, which enlarge the relations. These aspects bring new perspectives to the subject, such as: performing multiplications, additions, coordinating and dissociating actions with the intervention of external causes. Such questions, however, can not be resolved through concrete operations, requiring other, which are the formal operations.

Formal operational is the third stage, characterized by the release of the concrete, this means that "knowledge overpass the real to be inserted in the possible and to make a direct relation between the possibility and the necessity, without the fundamental mediation of the concrete"[24]. During this phase, the subject uses hypothesis (and not only objects), as well as propositions or relations among relations (second degree operations) and also performs operations such as inversion or negation, reciprocity and correlations.

4.2.2 Structures equilibration

After presenting the phases of cognitive structures development, we will discuss the construction of these structures, explaining the knowledge development and formation: Piaget presents the equilibration process as central, - states of qualitatively different equilibrium, with multiple disequilibration and reequilibration. For the development, the fundamental reequilibrations are those that promote a better equilibrium or a majorant equilibrium – the "*équilibration majorant*" – that leads to self-organization [21]. According to Piaget, it is important to analyze the causal mechanism of equilibrations and reequilibrations.

The concepts of assimiliation – integration of new elements into old structures or schemas and accommodation – modification of assimilation schemas as a result of external influence, are very important in this study. Knowledge is neither in the subject (innatism) nor in the object (empiricism), but it is constructed from an interaction between subject and object. When the subject acts upon the object, he extracts (abstracts) elements that are assimilated, this means, they are incorporated into a sensorial and conceptual schema through an interiorization process. This assimilation has to be accommodated, that means, it must take into consideration the singular features of the elements that were assimilated in order to allow their integration. The cognitive systems are then, at the same time, open systems (integration, assimilation) in what concerns the exchanges with the environment, and closed systems (differentiation, accommodation) as they are cycles. Equilibrium, therefore, is found through actions carried out in both directions. When there is an external disturbance, i.e. disequilibrium, there is a compensatory modification, an adaptation so that the cognitive system reaches a new equilibrium.

4.2.3 Disturbances, regulations and compensations

Piaget [21] also presents the regulation and compensation processes until the "*équilibration majorant*" is achieved: considering that the assimilation schemas bestow some meaning to assimilated objects, any obstacle to assimilation is seen as disturbances, and reactions to them are named regulations. Regulation consists of a recovery of an original action, modified by results - when the original action does not produce the expected results and it is taken again based on results obtained.

Disturbances may be of two types: (a) when there is an opposition to accommodation – the consciousness of such disturbance originates failure and error, and regulation is composed of a negative feedback: the correction; (b) when there are gaps – unfulfilled necessities, which refer to an object or necessary conditions to conclude an action or solve a problem. Such a blank is the assimilation schema that was activated, and regulation is composed of a positive feedback, as an extension of assimilation.

It is not possible to have regulation if there is only a repetition of an action, if there is not any change or if the action is interrupted. In other words, no regulation means none reequilibration. Regulations end up in compensations that can stabilize the initial actions, adding retroactive and proactive circuits, increasing the power of negotiations, or exceeding the initial actions towards a wider and more stable equilibrium. Both are constructive processes, even though they have different characteristics, as the latest leads to the possibility to understand new relations – "*équilibration majorant*". At short, cognitive equilibrations don't ever mark a stop point, because the equilibrium states are always surpassed; it is not only the case of a march to equilibrium but an entire structure oriented towards the best equilibrium.

4.2.4 Possibility and necessity

The reference to disturbances caused by the presence of blanks leads to studies about the *possibilities* – product of the subject's construction while interacting with the properties of the object, inserted in interpretations due to the subject's activities – and the *necessities* – product of the subject's inferential compositions [26].

Piaget found a relation between the levels of the operative stages and the formation of *possibilities*, which compose the relation between extrinsic and intrinsic variations: at the first level, new *possibilities* are opened step by step, successively, through retentions of a cognitive construction and small variations in it, which are updated. In this level, new (other) *possibilities* may be discovered through real experimentation, but they do not result on procedures that could lead to anticipations, because these *possibilities* initially are related to failures and hits, through partial or uncompleted laws. Only if corrected evaluations are joined with their necessities, then the *possibilities* can become deductible [25].

At the following level, *co-possibilities* (groups or families of procedures that complete the system of similarities and differences from the previous level) are anticipated by means of inferences. Such *co-possibilities* are concrete, initially, and they originate from the evolution of *possibilities* that were discovered in the previous level, and that become abstract due to the free imagination of previous *possibilities,* which need no more updates. This development of *possibilities* takes place through families or groups of procedures that complete the system of similarities and differences from the previous level.

The level of abstraction, which follows the two previous ones, accepts the existence of the infinitum, with the concept of the unlimited (unlimited combinations), and the understanding of anything (any combination). At this level, the subject's actions are not limited to the extrinsic and observable variations, as they are now supported by deductible, intrinsic variations. The operative structures appear, therefore, as a synthesis of the *possibilities* and *necessities* [25]. The *necessity* is the product of the subject's inferential compositions and, as well as the *possibility*, it is also unobservable [26].

4.2.5 Conducts, grasp of consciousness, action and concept

This research aims to apply Piaget's theory to education, especially to the study of pedagogical strategies which, as extrinsic factors, cause a cognitive disequilibrium and support student's intrinsic variables on the construction of an "équilibration majorant". Recapitulating the discussion on regulations, Piaget states that a regulation can be automatic - when there is little variation of means or little adjustments, or active - when the subject has to employ or choose other means. In the automatic regulation, there is not a grasp of consciousness, whereas the active regulation causes such a grasp [of consciousness], which leads to the representation or conceptualization of material actions [21].

In order to achieve a deeper understanding about such a discussion, it is important to define other two terms: *observables* and *coordination*. According to Piaget, *observable* is something that the subject perceives as a fact and, therefore, depends on instruments to register it (assimilation) through preoperational or operational schemas, which can modify data either through additional accuracy or through deformation. Because those schemas employ coordination, the observables are conditioned by previous coordination. There are observables which are perceived by the subject and observables which are registered on the object. Regarding the *coordination*, it is characterized by implicit or explicit inferences between the subjective evidence and the logical need. In this sense, it is not only a case of inductive generalizations, but the construction of new relationships - hypothesis - which are beyond the observable frontier [21].

Restating the concept of cognitive systems, they can be: *simple descriptions* – that means, observables which are conceptualized by the subject at the moment of the action or of the event; a *cognitive instrument* – employed by the subject in such conceptualizations (classifications, relationships); or *wider structures* – operational compositions, causal explanations, such as groupings, groups, etc.

Therefore, when there is a perturbation, the subject can reach the compensation through one of the possible cognitive conducts pointed out by Piaget:

Alpha conduct - the subject ignores or deforms the observables. If the disturb is too small, nearby the equilibrium point, the compensation occurs through a modification - in the inverse order of the disturbance. If the disturbance is more intense, the subject may cancel, neglect or dismiss the perturbation or, otherwise, take the perturbation into consideration, but with deformations. Therefore, such conducts are partially compensatory, and the resulted equilibrium remains very unstable.

Beta conduct - the subject modifies the assimilation schema; that is, he constructs new relationships. The disturbance element is integrated and modifies the system, and there is an equilibrium movement to assimilate the new fact. There is improvement, fusion, expansion, complementation or replacement through the construction of new relationships, that is, an internalization of the disturbances - which are turned into internal variations.

Gamma conduct - the subject anticipates the possible variations. When such variations become possible and deductible, they lose the disturbance characteristic and are inserted as virtual transformations of the system, which can be inferred. A new meaning is established and there is no longer compensation.

There is a systematic progress of these cognitive conducts - which comprise phases in accordance with the domains or raised problems - up to the level of the formal operations. In short, a characteristic of the *alpha* conduct is the lack of retroaction and anticipations. In this sense, *alpha* processes tend to cancel or dislocate the disturbances, while in *beta* conduct there is a possibility of a partial rearrangement or a more complete reorganization. *Gamma* conduct generalizes direct and inverse operative compositions, assimilating the perturbation.

From this standpoint, this study aims to model strategies that cause the student to reflect - in a more advanced stage, if possible - on the level of reflective abstraction with grasp of consciousness. The grasp of consciousness is a process that consists of a passage from the empirical assimilation (incorporation of an object into a scheme) to a conceptual assimilation [22, 23]. As a process, the grasp of consciousness is characterized by a consciousness *continuum* which begins in the action. The action is an autonomous knowledge susceptible of precocious achievement without grasp of consciousness.

In the first level, there are only material actions without conceptualization, because the subject uses empirical and pseudo-empirical abstractions for the regulation of new actions. There is an internalization of the actions through assimilation of the schemas, and the externalization occurs through the accommodation of the subject - either through the orientation of instrumental behaviors or the logic of actions. Such actions are automatized or learned actions that are not always understood or susceptible to be conceptualized by the subject.

In a second level, the appropriation of the coordination mechanisms of the actions enables the construction of the operations - comprising the grasp of consciousness. The internalization is carried out by an empirical abstraction, as well as by a reflective abstraction - and it is externalized by the representation of the observed data, deductive interpretations and causal explanations. Such phase is normally long. The action and the conceptualization are approximately at the same level, in a constant dialectic. It means that the subject may employ plans (although they are restricted) or make choices based on his/her conceptualizations.

The accomplishment of new operations over previous operations - carried out through reflective abstractions, enables the conceptualization to overtake the action and, therefore, to guide it. Such process characterizes the third level, in which the internalization occurs through reflective abstractions - which are externalized by the possibility of varying the factors, the experimentation, and the construction of models or hypothesis.

Summing up, a sensorimotor achievement does not imply that a grasp of consciousness has happened, although it always occurs from an action in elementary situations. Notwithstanding, from a certain level and in more complex situations, there are influences resulted from the concept guiding the action. If in elementary situations it was possible *to do something without understanding it*, upon *the reflection about the doing of something,* compensation normally occurs. On the other hand, in higher levels, it is possible to *think and to experiment how to do or create a different way of doing something.*

4.3 AMPLIA

The objective of AMPLIA – an Intelligent Multi-agent Probabilistic Learning Environment [33] is to be a qualified additional resource to medical education, supporting diagnostic reasoning development and modeling of diagnostic hypotheses. It is composed of a multi-agent system and BNs [28], which have been widely used to model uncertain domains [14] such as medicine.

Uncertainty is represented through probabilities and the basic inference is the probabilistic reasoning, this means, the probability of one or more variables undertaking specific values considering the available evidence. Pearl [20] pointed out some empiric evidence that the probabilistic reasoning is similar to the human reasoning patterns. Yet, the hypothesis that physicians implicitly perform a probabilistic reasoning when making diagnoses is supported by reviews of case studies in the medical field [11, 16, 28].

The proposal of AMPLIA is to offer an open environment where a student can build a graphical model that represents his diagnostic hypothesis for a clinical case, using BNs. The student network is compared to that of an expert on the domain stored within the environment, and the differences among such networks are managed by an intelligent agent that uses pedagogic strategies based on the constructivist model.

AMPLIA is formed by three types of cognitive agents (Learner Agent, Mediator Agent and Domain Agent) that communicate through a server (the ComServer), as Fig. 4.1 shows. The environment still counts on two database (Domain DB) and (Pedagogic Strategies DB) and an interface with the BN editor, named SEAMED [7].

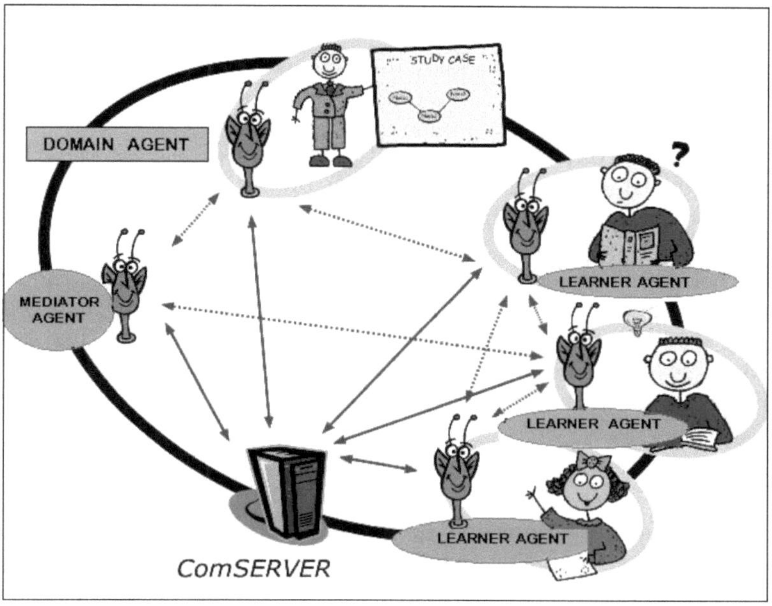

Fig. 4.1. The intelligent agents of AMPLIA

4.3.1 Domain Agent

The responsibility of the Domain Agent is to evaluate the network the student builds by comparing it against the expert's network as to the items: feasibility, correctness and completeness. The agent performs two evaluation processes [9]: a qualitative assessment (to analyze the network topology) and a quantitative assessment (to analyze the tables of conditional probabilities). The topology is analyzed by (a) a list of variables specific for each case, (b) the expectation of the type of inference the student makes and (c) simplification of the expert's network, according to the case the student selected. The parameters of the list of variables are: the name of the node and classification as to function and importance of the expert's network. (See Table 4.1).

When the student network reaches a good probability of being in a satisfactory level in the qualitative assessment, the Domain Agent starts to analyze

Table 4.1. Classification of the nodes of the expert's BN

Diagnostic	Case solution diagnosis
Trigger	Selects the diagnosis as potential solution to the problem
Essential	Warrants the diagnosis identification
Complementary	Increases the diagnosis probability
Excluding	Shows the diagnosis is not probable, i.e., it has low probability
Unnecessary	Totally unnecessary to diagnosis

the distribution of conditional probabilities among variables as well, thus performing a quantitative evaluation. In such process, the student network is submitted to a database with real cases, so that its performance can be assessed and the possibility of leading to a correct diagnosis is also checked. The Learner Agent represents the student's belief on the domain (the student network) and the confidence level he has on his own network. This Agent also infers the student's autonomy, based on the student's actions. The outcome of the analysis described is sent to the Mediator Agent, which is responsible for the selection of pedagogic strategies.

4.3.2 Learner Agent

The Learner Agent is responsible for the construction of the student model, by observing his actions in a graphic editor. Such actions may be observed through the log and they are composed of the process of adding (*inserção*) and removing (*exclusão*) arcs (*seta*) and nodes (*nodos*) while the student builds his BN. Figure 4.2 show a sample of the log recorded by the Learner Agent. This student inserted four nodes and then four arcs, then one more node, one more arc, removed this arc, removed the last node, added three other nodes, removed one of them, and so on. As much as the student insert and remove nodes and/or arcs, more the Learner Agent will infer, in a probabilistic way, that the student is constructing his network by means of trials and not based on a hypothesis. The Learner Agent will classify this student, at this moment, with a low credibility.

The Learner Agent has a mathematical approach to such process [30] and obtains the first (*a priori*) probabilities for the variables that will compose the nodes (Nodes, Arcs, Network) of a BN, as shown in Fig. 4.3. It will infer the level of credibility that the agent may have towards the student's autonomy while accomplishing his tasks (Credibility). The nodes Diagnosis and Unnecessary are directly observed in the log.

In a pedagogic context, the Learner Agent represents the student by constructing a model of this student, while the teacher role is divided into two other agents: the Domain Agent as the expert on the domain, and the Mediator Agent as the responsible for the process of pedagogic negotiation. This

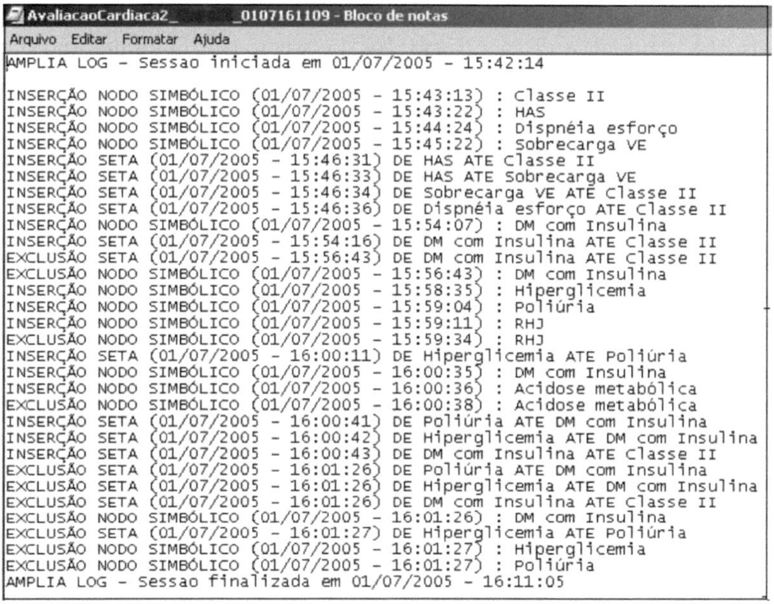

AvaliacaoCardiaca2_[redacted]_0107161109 - Bloco de notas
Arquivo Editar Formatar Ajuda

```
AMPLIA LOG - Sessao iniciada em 01/07/2005 - 15:42:14

INSERÇÃO NODO SIMBÓLICO (01/07/2005 - 15:43:13) : Classe II
INSERÇÃO NODO SIMBÓLICO (01/07/2005 - 15:43:22) : HAS
INSERÇÃO NODO SIMBÓLICO (01/07/2005 - 15:44:24) : Dispnéia esforço
INSERÇÃO NODO SIMBÓLICO (01/07/2005 - 15:45:22) : Sobrecarga VE
INSERÇÃO SETA (01/07/2005 - 15:46:31) DE HAS ATE Classe II
INSERÇÃO SETA (01/07/2005 - 15:46:33) DE HAS ATE Sobrecarga VE
INSERÇÃO SETA (01/07/2005 - 15:46:34) DE Sobrecarga VE ATE Classe II
INSERÇÃO SETA (01/07/2005 - 15:46:36) DE Dispnéia esforço ATE Classe II
INSERÇÃO NODO SIMBÓLICO (01/07/2005 - 15:54:07) : DM com Insulina
INSERÇÃO SETA (01/07/2005 - 15:54:16) DE DM com Insulina ATE Classe II
EXCLUSÃO SETA (01/07/2005 - 15:56:43) DE DM com Insulina ATE Classe II
EXCLUSÃO NODO SIMBÓLICO (01/07/2005 - 15:56:43) : DM com Insulina
INSERÇÃO NODO SIMBÓLICO (01/07/2005 - 15:58:35) : Hiperglicemia
INSERÇÃO NODO SIMBÓLICO (01/07/2005 - 15:59:04) : Poliúria
INSERÇÃO NODO SIMBÓLICO (01/07/2005 - 15:59:11) : RHJ
EXCLUSÃO NODO SIMBÓLICO (01/07/2005 - 15:59:34) : RHJ
INSERÇÃO SETA (01/07/2005 - 16:00:11) DE Hiperglicemia ATE Poliúria
INSERÇÃO NODO SIMBÓLICO (01/07/2005 - 16:00:35) : DM com Insulina
INSERÇÃO NODO SIMBÓLICO (01/07/2005 - 16:00:36) : Acidose metabólica
EXCLUSÃO NODO SIMBÓLICO (01/07/2005 - 16:00:38) : Acidose metabólica
INSERÇÃO SETA (01/07/2005 - 16:00:41) DE Poliúria ATE DM com Insulina
INSERÇÃO SETA (01/07/2005 - 16:00:42) DE Hiperglicemia ATE DM com Insulina
INSERÇÃO SETA (01/07/2005 - 16:00:43) DE DM com Insulina ATE Classe II
EXCLUSÃO SETA (01/07/2005 - 16:01:26) DE Poliúria ATE DM com Insulina
EXCLUSÃO SETA (01/07/2005 - 16:01:26) DE Hiperglicemia ATE DM com Insulina
EXCLUSÃO SETA (01/07/2005 - 16:01:26) DE DM com Insulina ATE Classe II
EXCLUSÃO NODO SIMBÓLICO (01/07/2005 - 16:01:26) : DM com Insulina
EXCLUSÃO SETA (01/07/2005 - 16:01:27) DE Hiperglicemia ATE Poliúria
EXCLUSÃO NODO SIMBÓLICO (01/07/2005 - 16:01:27) : Hiperglicemia
EXCLUSÃO NODO SIMBÓLICO (01/07/2005 - 16:01:27) : Poliúria
AMPLIA LOG - Sessao finalizada em 01/07/2005 - 16:11:05
```

Fig. 4.2. Sample of a log recorded by the Learner Agent

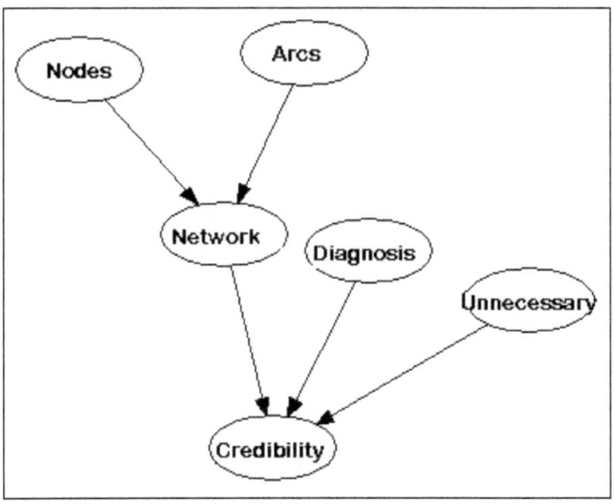

Fig. 4.3. BN of the Learner Agent

process aims at solving conflicts that may occur in the teacher's evaluation towards the student and vice-versa (or between the Domain and Learner Agents). It uses argumentation mechanisms that aim at strengthening the individual and mutual confidence of the participants with relation to the domain approached [20]. Arguments used in the negotiation carried out by the Mediator Agent compose the pedagogic strategies, which will be further discussed in the next session.

4.4 Pedagogic Strategies

4.4.1 Creating strategies and tactics

When the student starts a study session in AMPLIA, he selects a clinical case and he receives a text with information such as patient's history, anamnesis, laboratory tests, etc. Then, the student accesses the screen with tools for nodes selection and arcs insertion. Nodes contain information on the case and arcs indicate the dependence relations among, directed from the parent nodes to the child nodes, so that the child node is influenced by its parents. Figure 4.4 shows the user's screen in a study session.

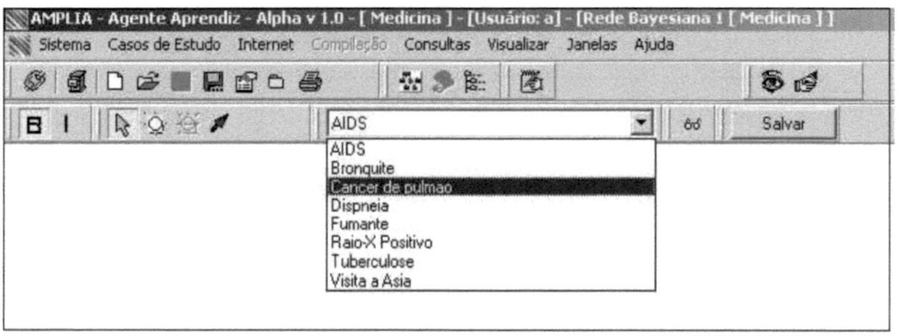

Fig. 4.4. User's screen

The network expresses the diagnostic hypothesis graphically, through the topology, while the intensity of the dependence relations is informed through the tables of conditional probabilities, available for each node. When the BN is executed, it propagates such probabilities, supplying the final (*a posteriori*) probabilities for each node. When new evidence is entered, these probabilities change. Figure 4.5 shows an example of a Bayesian network built by a student, and the scale for the self-confidence declaration, which the student declares every time he submits his BN to AMPLIA

In this example, the study case is about cancer, bronchitis and tuberculosis diagnose, (the last one is not represented) and the student has to find out the correct symptoms (only smoker – fumante – is represented), lab tests and

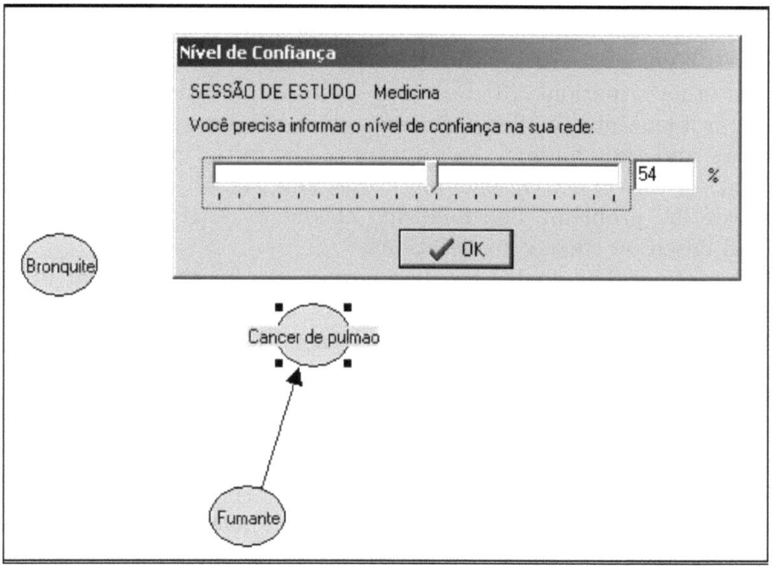

Fig. 4.5. Example of BN built by a student and self-confidence declaration

their probabilities, in order to justify his hypothesis, this means that he has a previous hypothesis and he just confirm it using the data – this is called diagnostic reasoning.

This virtual data manipulation consists itself, a powerful strategy, as the student can "see" in a concrete way all the possible combinations among the data and he can test all the different probabilities. As seen, it is important for the student to act over the object (to construct his network), to realize empirical abstractions (to observe the outcome of his actions), to coordinate this actions by means of inferences (reflective abstraction) and to reflect about this abstractions, elaborating hypothesis that can lead to new actions which can confirm them (or not).

So, strategies used in AMPLIA are aimed to make the student conscious about a study case, so that he is able to make reflective abstractions about the case and, if possible, reflected abstractions. The expectation is that the Mediator Agent causes a cognitive disequilibration in the student, followed by an "*équilibration majorant*", therefore strategies are elaborated considering: (a) level of the student's grasp of consciousness, inferred from the major problem detected by the Domain Agent in the student's network and his declaration of self-confidence, and (b) level of the student's action autonomy, inferred by the Learner Agent.

In Action and Concept [22] Piaget analyzed the different actions subjects take to solve a problem, and studied these processes and the relation between observables and the coordination of the actions. He detected the following procedures: (1) Space opening: creation of new ways of performing some procedure; (2) Conservation: repetition – using a known and well-succeed procedure in other situations; (3) Decomposition of familiar schemas: to decompose the problem into smaller problems, which can be solved by using known procedures and thus finding the solution of the initial problem; (4) "Transformation" of the object (re-meaning): assigning the object another meaning that solves the problem. The following classes of AMPLIA strategies were organized based on these considerations:

- *Orientation:* The goal is to open new spaces for the student in case he builds a network that is not feasible. Direct information is provided to the student so that he can build the network in a different manner (so that his network becomes a BN).

- *Support*: This strategy also presents concrete and contextualized data, so that the student can increase his confidence;

- *Contest:* This strategy aims at warning the user about inconsistencies in his network, fostering a new assessment, so that the student can redo some procedures based on those that presented good results. The procedures of conservation and decomposition are involved in the *Contest* strategy.

- *Confirmation and Widening*: these strategies are directed towards the third level of grasp of consciousness, requiring reflected abstractions, such as the variation of experimentation factors and construction of new hypothesis. *Confirmation* takes place through the presentation of data and hypotheses, which aim at making the student, reflect and increase his self-confidence, and *Widening* aims at stimulating the production of new hypotheses.

As a *Strategy* is understood as a plan, construction or elaboration, the action is then named *Tactics*. In other words, strategy is a cognitive process that aims at reaching an objective, but that must be accomplished, performed or, in this case, presented to the student through a tactics. The creation of tactics for the display of strategies takes into account the student's autonomy level, with basis on the inference of credibility the Learner Agent makes. It evaluates whether the student's actions are more guided by the objects observables or by his reflections.

The studies by Inhelder and Cellérier [12] identified two types of procedures: a) *Bottom up*, which is characterized by concrete actions, as when the student has the information and looks for the hypothesis, and b) *Top down*, with predominance of cognitive actions, when the student has a hypothesis and looks for the information with enhanced autonomy.

Considering that the presentation of these tactics is, overall, a dialogic relation among different parts, a rhetoric condition is involved. By analyzing the understanding of rhetoric in university students, Cubo de Severino [5] studied the cognitive processes that it comprises, in an increasing scale of abstraction levels, from examples up to general principles. She describes six levels: a) narration - follows the concrete experience; b) examples – recover common sense concepts; c) comparison – uses category paradigms or prototypes and inferences; d) generalization or specification – uses models of known knowledge, and e) classification – relations that may allow classification, and f) definition – identifies new concepts. We observe then the importance of rhetoric to adequate communication to individual features: ascending, when actions take place in the direction concrete to abstract, and descending, when they are in the opposite direction.

Thus, each strategy class previously elaborated was added basic tactics, requiring different levels of student's autonomy:

- Orientation strategy: tactics are *Correction* (to redo the action), *Indication* (to follow some indication) and *Suggestion* (to decide, based on concrete material).
- Contest strategy: tactics are Experimentation (possibility of manipulating data), Search (search data) and Reflection (analyze data);
- Support strategy: uses the *Example* tactics, which is "to do the same";
- Validation strategy: tactics are: *Demonstration,* which means the observation of data presented, and the expert's *Model Presentation*;
- Widening strategy: tactics are *Discussion* (using an argumentative text), *Hypothesis* (formulation a hypothesis to anticipate an action) and *Problematization*, which is the data organization.

4.4.2 Tactics selection

Tactics are selected by the Mediator Agent [30, 31], through an Influence Diagram[5] (ID), as shown in Fig. 4.6. This ID is an amplified BN, therefore, it is probabilistic as well. It helps decision taking based on the utility function (*Utility*), according to the student's autonomy level – the selected strategy is expected to be the most useful for the student, on that moment of his process for knowledge construction. Variables considered for tactics selection are: *major problem* (*Main Problem*) found in the student's BN, which is informed by the Domain Agent, and Mediator Agent then evaluates the most likely classification for the student's network (*Learner Network*) (see Table 4.2), *confidence level* (*Confidence*) the student declares and the *level of credibility on the student's autonomy* (*Credibility*) result from the inference by the Learner Agent.

The process of tactics selection can be retrieved from the ID database in the .xml format. Figure 4.7 shows the aspect of an ID file of the Mediator Agent. In this example, the Main Problem was a disconnected node

[5] Influence Diagram (ID) is an acyclic oriented graph with nodes and arcs: *Probability* nodes, random variables (oval). Each node has an associated table of conditional probabilities; *Decision* nodes, points of actions choice (rectangles). Its parents nodes can be other decision nodes or probability nodes; *Utility* nodes, utility functions (lozenges). Each node has a table containing the description of the utility function of variables associated to its parents. Its parents can be Decision nodes or Probability nodes. Conditional arcs arrive on probabilistic or utility nodes and represent the probabilistic dependence.

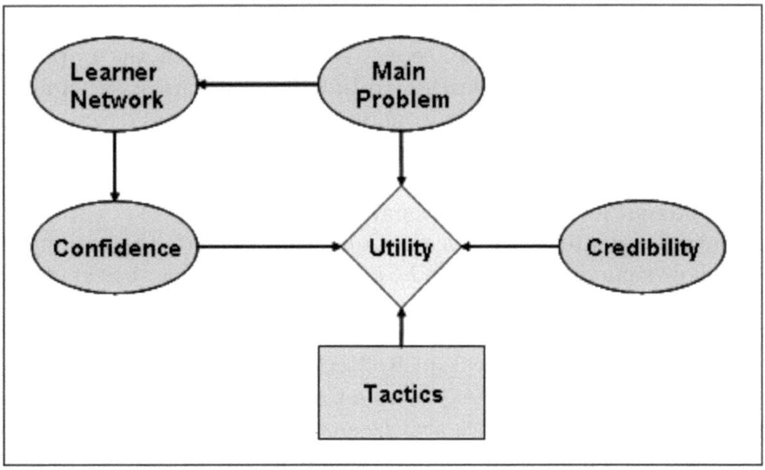

Fig. 4.6. Influence Diagram of the Mediator Agent

Table 4.2. Classification of the BN (*Learner Network* node) according to the major problem

Not feasible (INV)	Presence of cycles or disconnected nodes (is not a BN)
Incorrect (INC)	Absence of diagnostic node, presence of diagnostic node as parent node of symptoms, presence of excluding node with incorrect representation of probabilities
Potential (POT)	Presence of trigger nodes only and/or essential and/or unnecessary nodes, besides the diagnostic nodes
Satisfying (SATISF)	Absence of complementary nodes and/or arcs
Complete (COMPL)	Network without errors and with a good performance as compared to the expert's model and to the database of real cases

(*no_desconex*), so the Learner Network is classified as Not feasible (*inviável*); the student declared his self confidence Low (*baixa*) and the inferred credibility was High (*Alto*). With this information, the Mediator Agent decided to use a Suggestion (*sugestão*) to present the Orientation strategy – the student will receive a message like this: "*your network doesn't correspond to a BN; you can read a text about the matter*", followed by the text. If the confidence and/or credibility were different, the messages could be some others, like "*your network doesn't correspond to a BN; here is an example of a BN*", or "*There are disconnected nodes in your network*", followed by a highlight of the disconnected node.

The tactics are presented to the student as messages on the screen, the student can read them and, if necessary, he can access supplementary information like text, papers or links and modify his network. Here are some examples

```
<user[redacted] hora="14:12:32" IdExercicio="AvaliacaoCardiaca1-858-070705-140550">
<ciclica>0</ciclica>
<f_trigger>0</f_trigger>
<f_essenciais>0</f_essenciais>
<f_complemen>0.032735162909027</f_complemen>
<no_desneces>0</no_desneces>
<s_problemas>0</s_problemas>
<no_desconex>0.960591371296005</no_desconex> <– Main Problem
<no_exclud_er>0</no_exclud_er>
<diagnosti_pai>0</diagnosti_pai>
<s_diagnostic>0</s_diagnostic>
<f_arcos>0.00667346579496792</f_arcos>
<potencial>0.0299102925314855</potencial>
<completa>0.000274252018971285</completa>
<satisfatoria>0.0092240841535382</satisfatoria>
<inviável>0.960591371296005</inviável> <– Learner Network
<incorreta>0</incorreta>
<baixa>1</baixa> <– Confidence
<media>0</media>
<alta>0</alta>
<Baixo>0</Baixo>
<Medio>0</Medio>
<Alto>1</Alto> <– Credibility
<correcao>-0.00917427462536195</correcao>
<indicacao>-0.00917427462536195</indicacao>
<sugestao>2.56904708537464</sugestao> <– Tactic of orientation strategy
<experimentac>-0.00917427462536195</experimentac>
<busca>-0.00917427462536195</busca>
<reflexao>-0.00917427462536195</reflexao>
<exemplos>-0.00917427462536195</exemplos>
<problematizac>0.0103503868967181</problematizac>
<discussao>-0.00917427462536195</discussao>
<demonstraca>-0.00917427462536195</demonstraca>
<hipoteses>0.000776581674638053</hipoteses>
<mod_esp>-0.00917427462536195</mod_esp>
</user[redacted]>
```


Fig. 4.7. ID file of the Mediator Agent

of screen shots for different tactics, employed in an experimental clinical test, about differentiating the diagnosis of bronchitis, cancer and tuberculosis.

Figure 4.8 shows a not feasible network as there is a disconnected node (bronquite) at left. The selected strategy is orientation (as seen previously), presented by means of the message "*Your network is not in accordance to the concept of BN. Here you can see the definition and examples of a BN*". This is an indication strategy, because the Learner Agent inferred a low credibility level for this student meaning that the student probably needs a more objective help, i.e, to follow some indication.

Figure 4.9 shows an incorrect network, because, in this study case, there are three diagnoses nodes (*tuberculosis* is missing). As the Learner Agent inferred a medium credibility and the student declared a low self confidence, the selected strategy is to contest this network using a searching tactic: "*An important node is missing in your network. Please, verify which diagnosis is justifying your hypothesis*". In this case, as the student probably has a hypotheses, but he is not well confident about it, the agent will wait for the student's action, (like inserting a node), in order to give him a "chance".

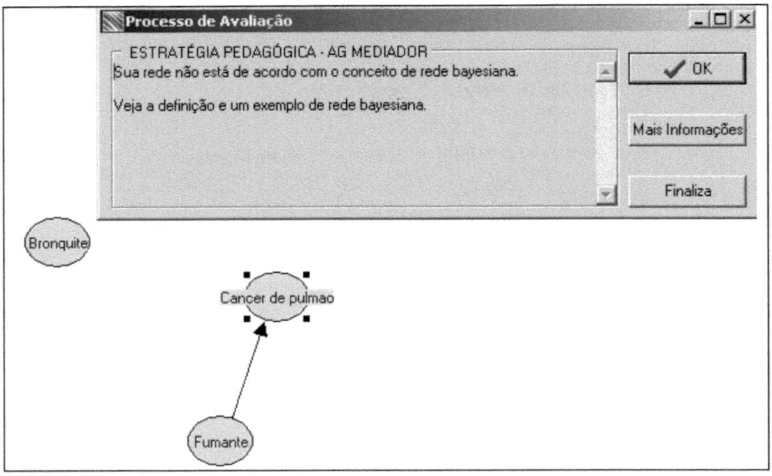

Fig. 4.8. Tactic for a not feasible network

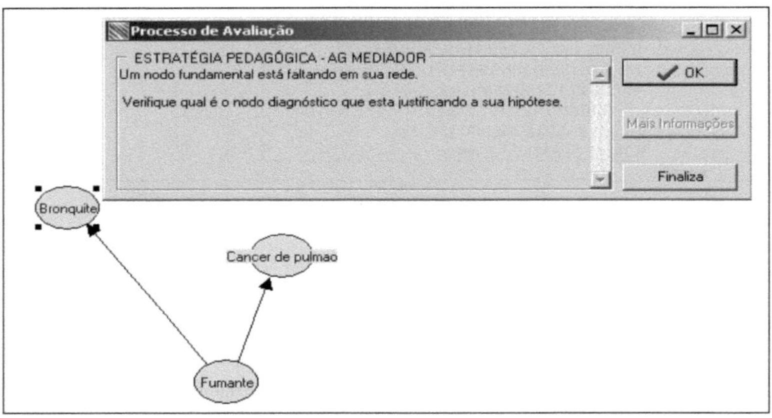

Fig. 4.9. Tactic for an incorrect network

Figure 4.10 shows an example of the strategy and tactic for a satisfactory network – widening using a discussion argument. This network is almost complete, only complementary information is missing, and also some conditional probabilities are not informed. The student declared a high self confidence level, and his credibility was also inferred in a high level. This student will receive a message like this: "*Your network is satisfactory, but you have to verify the conditional probabilities table. There are also some complementary nodes missing in your network, you can find that out in these additional sources*". In this case, the student will receive internet links, selected by AMPLIA, so he can search for new information about the study case.

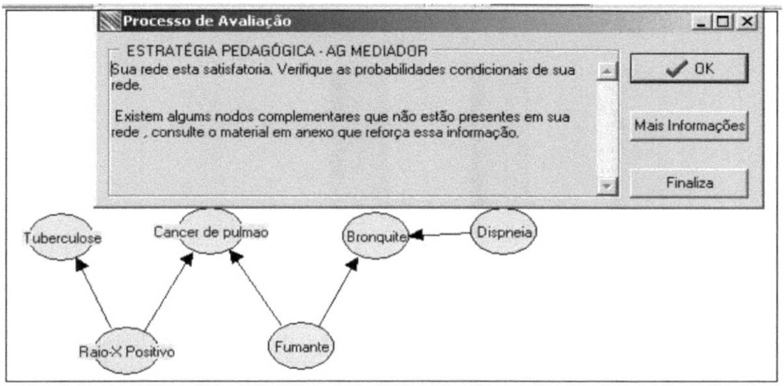

Fig. 4.10. Tactic for a satisfactory network

4.5 Tracking the use of strategies in AMPLIA

A prototype of AMPLIA with a network about Congestive Cardiac Failure was implemented. The network was built by an expert on the domain, which allowed the first tests in real conditions to be carried out. This first experimental phase counted on seven medical students and eleven resident physicians from Hospital de Clínicas de Porto Alegre, who attended study sessions from May to June 2005 [31]. Studies for statistics purposes and more refined conclusions are still under development.

This section shows an example of one of those study sessions, where we can follow how the quality of the network increases and how the confidence and credibility levels vary along the process.

AMPLIA allows for the online observation of the negotiation process, which consists of interaction cycles between the student and the Mediator Agent, through a dynamic virtual graphic made available by the Mediator Agent. The analysis of the Learner Agent log records with the student's actions in the .txt format, and the .xml files of the Mediator Agent database provides information about the student's cognitive status, and about the impact of strategies used, thus making possible to track the process.

The Figs. 4.11 and 4.12 present twelve interaction cycles. In each cycle of the interaction, defined by the BN model submission, the left column indicates the S*elfconfidence*, and the right column indicates the inferred *Credibility*. These values can be: Low (*Baixo/a*), Medium (*Médio/a*) or High (*Alto/a*). The horizontal line represents the BN model *classification*, from the lowest (Not feasible) to the highest point (Complete). The student's log files, as well as the Mediator Agent files are not attached to this article for being too extensive, but the reader may find them by consulting the references [31].

Cycle 1 – The student inserted many nodes and arcs, removing and replacing some of them. As his network was not feasible and the confidence declared was low, his most probable cognitive conduct is of *alpha* type, this

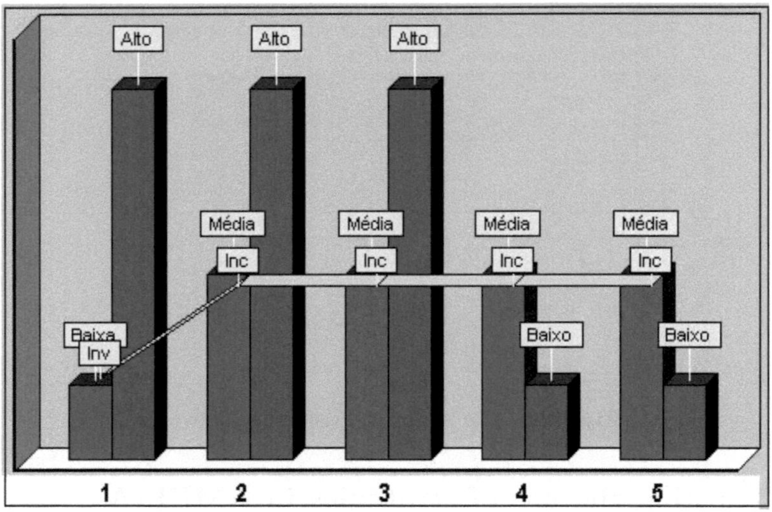

Fig. 4.11. Interactions performed by Student A (cycles from 1 to 5)

means, observables are ignored. In fact, the Domain Agent evaluation detected the existence of disconnected nodes. The strategy used in this case was to *guide* the student by suggesting the use of a simple material, so that the student could verify on his own why his network was not feasible.

Cycle 2 – The student inserted an arrow (arc) to connect the disconnected node and declared medium confidence; therefore there was an increase in the confidence level as compared to the previous cycle. The major problem detected now is the existence of a "parent diagnose" - there are arcs from diagnosis towards evidence. The Mediator Agent strategy is to contest the student network and it sends the reflection tactics through a message such as: "Your network is not oriented towards a diagnostic hypothesis. Look again and built your network so that the symptoms justify the diagnosis."

Cycle 3 – The log shows that the student submitted his network again within an interval of 19" without any changes. The reflection tactic is sent again. Only at this moment the student performed some changes in his network, as the next cycle shows.

Cycle 4 – The student excluded almost all arcs, replaced then in the opposite direction and this decreased his credibility. In this cycle, we see that the student has changed his network by making trials, probably guided more by the observables than by his hypotheses, in a bottom-up procedure, which made the Learner Agent infer a low credibility as to the student's autonomy. As the network was still incorrect, the tactic employed was *experimentation*, which is the possibility of manipulating data in a concrete way, through the presentation of an example and an invitation to build the network according to the example.

Cycle 5 – Another arrow was changed, but the evaluation shows that there still are wrong arcs, the tactics is *experimentation* again. The same process is repeated in this cycle.

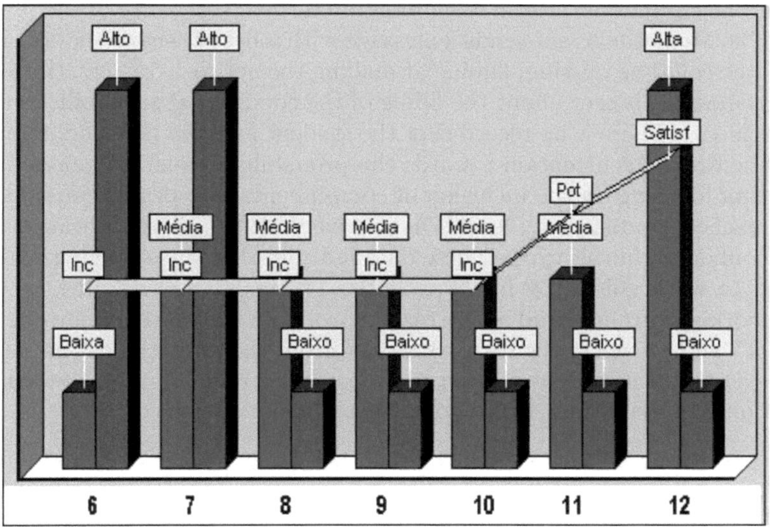

Fig. 4.12. Interactions performed by Student A (cycles from 6 to 12)

Cycle 6 – The student declared a low confidence, probably because he is disturbed for not being able to correct his network, or at least for not becoming conscious of that, because the inference of the Learner Agent indicates an increase of credibility, when the student adequately excluded an arc. The tactics the Mediator Agent sends is *search* (search the arcs connected to the diagnosis), because the network still has problems related to the "parent diagnose"), but the Agent tries to make the student increase his confidence (and became aware) of his actions.

Cycle 7 – In this cycle, we see an increase of confidence (medium confidence) but the result of the evaluation indicates that there still are inverted arcs. The student now inserts a new arc and excludes the other. The tactics is an invitation to reflect about the fact, making possible the student reflects about his actions.

Cycles 8 to 10 – We clearly observe that the student is making conservative and decomposition procedures, because in this cycle he changed one arc at a time and evaluated his network, which maintained the previous classification. The student's credibility decreased and the tactic used was experimentation. AMPLIA tries to cause a cognitive disequilibrium in the student, but he acts step-by-step, through small concrete changes, which are typical in the *alpha* conduct.

Cycle 11 – In this cycle the network evaluation reaches the potential level, because the major problem is now the lack of complimentary nodes. The system offers a strategy to widen the hypothesis through problematization about nodes, counting on *beta* conduct from the student - this means he is able to build new relations, after having integrated the disturbance of the previous level. The Mediator Agent sends a message with a list of random nodes, asking which one could be missing, aiming at making the network organization easier. A warn message is sent about the filling of the conditional probabilities table.

Cycle 12 – There is no record that the student inserted new nodes and certainly he drove his attention towards the probabilistic relation among nodes, instead of focusing on the inclusion of complimentary nodes, because his network reached a satisfactory level. We observe that the student declared maximum confidence in this cycle. These data indicate that the student apparently started to work guided by his hypothesis. Therefore, if the student was initially working through trial and error, he probably became conscious between cycles 11 and 12, when the network passed from the potential to the satisfactory level and the student started to reflect upon relations and to work with conditional probabilities, which constitute a *gamma*-type conduct.

4.6 AMPLIA as medical learning software

As AMPLIA purpose is to be a constructivist learning environment (CLE), in this section we will briefly discuss these environments and medical learning software, highlighting AMPLIA unique features and how it fits to those categories. According to Jonassen [15], CLEs must have some features in common, such as: capability to represent the natural complexity of the real world, focus on knowledge construction instead of reproduction, authentic activities, which are contextualized and not only composed of abstract instructions; learning based on real cases and not in pre-programmed sequences; they must stimulate reflection, relations between content and context and support to collaborative construction through social negotiations.

Similarly, Murphy [18] also presents a relation of several items that define a constructivist environment, such as: making possible to represent multiple perspectives, to have student-oriented objectives, the teacher acting as a mediator, fostering of metacognition, student's control over his own process allowed; authentic activities and contexts, knowledge construction and sharing allowed; valuing of previous knowledge; management of problem solving; mistakes taken into account, work with concepts interrelationships, etc.

According to Jonassen, the learner has to be active, to manipulate objects, to integrate new concepts, to build models, to explain things, and to collaborate to each other. They also have to be reflective and critical. Learning activities must be set within a context, featuring the real world complexity and with objectives that are clear to learners. Strategies (or instructional processes) the author mentions are the use of models (demonstrations or examples); tutorship (which means offering help when required) and fitting of activities according to the student's performance level.

Provided that AMPLIA is a learning environment with pedagogic strategies based on the student's cognitive model, following a constructivist approach, and which uses Bayesian networks to represent knowledge, we highlight some learning environments and medical software that can be used for education purposes. Such systems, further detailed in the comparative Table 4.3, contain one or more of those features mentioned.

As shown in this table, the main focus and the unique feature of AMPLIA, as compared to the others, is that it takes into account the cognitive state to build the student model, following an epistemological theory. Most systems use models based on knowledge and one of them also takes into account self-confidence. Strategies used in the systems compared above do not account for the student's cognitive state. Thus, we believe AMPLIA contributes to the development of CLEs, based on the studies by Piaget [24] and the genetic epistemology.

Table 4.3. Comparison among systems employed in medical education

Systems	Aims to	User's Interface	Student's actions	Student's model	Strategies
AMPLIA	Diagnostic hypothesis construction	Bayesian networks	Constructing and testing hypothesis	Knowledge; Self confidence; Cognitive state	From hints and quizzes to problems and discussions
Bio World	Problem solving	Text Frames Multimidia	Constructing hypothesis	Knowledge Self confidence	Contextual help
COMET	Problem solving Collaborative learning	Bayesian networks, Chat, Figures	Constructing hypothesis; Searching for medical concepts	Individual and groups Knowledge and Activities	From hints to collaborative discussion
Medikus	Problem solving	Bayesian networks	Constructing his model	Knowledge	Help Suggestions
Promedas	Diagnostic decision support	Bayesian networks	Entering findings	Knowledge	Explanations

4.7 Summary

AMPLIA was developed as a learning environment for the medical area, based on the constructivist theory by Piaget. The environment allows students to build a representation of their diagnostic hypotheses and to train their clinical reasoning, with the aid of pedagogic strategies that take into consideration the student's cognitive conduct.

The clinical reasoning is the way how an expert resolves a clinical case – from a possible diagnostic hypothesis the professionals look for evidence that confirm or reject his hypothesis. This type of reasoning is named top-down, because it starts from the diagnostic to find evidence, this way, the evidence justify the diagnostic. The student, however, makes the contrary, he looks for a diagnostic that justify the evidence, because he does not have a diagnostic hypothesis. His reasoning is bottom-up, starting from evidence to reach a diagnostic. We highlight the pedagogic function of constructing the diagnostic reasoning as an important cognitive process for the understanding of the clinical procedure based on Piaget's studies. By the grasp of consciousness, the subject's actions are guided by concepts, models and hypothesis.

AMPLIA's pedagogic strategies were defined after many considerations about the process of knowledge building and subject's regulatory conducts in the equilibration process, highlighting the "équilibration majorant", within the process of grasp of consciousness.

These strategies were considered as arguments within a process of pedagogic negotiation between the intelligent agents of AMPLIA and the student. When an intelligent agent was assumed to be the mediator of this process, variables had to be selected for the construction of the student model and of the pedagogic agent. For that end, cognitive, affective, procedural and domain aspects were considered. They were considered uncertain knowledge represented through probabilistic networks. Such variables allow the Mediator Agent to update its arguments according to the student's cognitive state, so that the student interacts with the environment through different strategies along the process. For example, if the student's conduct is to use concrete actions, without retroaction and relationships, which are typical of an *alpha* conduct, the tactics the intelligent agent will use will also be directed towards this level of action. It is important to mention, however, that not every action lead to a conscious awareness, because there are stages or cognitive levels in which the subject's action are well succeeded, even if there is not a grasp of consciousness. Similarly, as the interaction between the student and the Mediator Agent is probabilistic, the student may have a cognitive disturbance, even if he is not interacting by means of a strategy specifically turned towards the objective of an "équilibration majorant".

The example presented was aimed at demonstrating the interaction cycles between a student and the Mediator Agent. We could observe the gradual improvement of the network quality, and the possibility of an analysis of different cognitive conduct, from *alpha* to *gamma* conduct, in the beginning of reflection about the conditional probabilities.

We conclude, therefore, that intelligent agents can use pedagogic strategies based on the student's cognitive model, which is also built by intelligent agents. Such feature makes AMPLIA a powerful model, which contributes to the development of CLEs with a basis on Piaget's studies [24] and the genetic epistemology.

Acknowledgments. This work received financial support from CAPES, Brazilian Agency which gives support to human resources qualification.

References

1. Baylor AL (1999) Intelligent agents as cognitive tools for education. Educational Technology 39:36-40
2. Bercht M (2000) Pedagogical agents with affective and cognitive dimensions. In: Actas RIBIE. Universidad de Chile, Santiago Chile
3. Bull S, Pain H (1995) Did I say what I think I said, and do you agree with me?: inspecting and questioning the student model. In: Greer J (ed) Proceedings AIED'95. AACE, Washington DC 269-27
4. Conati C et al (1997) On-line student modeling for coached problem solving using Bayesian networks. In: Proceedings UM97. Springer, Vienna 231-42
5. Cubo de Severino L (2002) Evaluación de estrategias retóricas en la comprensión de manuales universitários. Revista del instituto de investigaciones linguisticas y literárias hispanoamericanas RILL 15
6. Fenton-Kerr T (1999) Pedagogical software agents. Synergy 10:144
7. Flores CD, Ladeira M, Viccari RM, Höher CL (2001) Una experiencia en el uso de redes probabilísticas en el diagnóstico médico. Informatica Medica 8:25-29
8. Flores CD (2005) Negociação pedagógica aplicada a um ambiente multiagente de aprendizagem colaborativa. Phd Thesis, UFRGS, Porto Alegre Brazil
9. Gluz JC (2005) Formalização da comunicação de conhecimentos probabilísticos em sistemas multiagentes: uma abordagem baseada em lógica probabilística. Phd Thesis, UFRGS, Porto Alegre Brazil
10. Halff HM (1988) Curriculum and instruction in automated tutors. In: Olson MC, Richardson JJ Intelligent Tutoring Systems. Lawrence Erlbaum, London
11. Heckerman D, Horvitz E, Nathwani B (1992) Towards normative experts systems: Part I. The Pathfinder project. Methods of Information in Medicine 31:90-105
12. Inhelder B, Cellérier G (1996) O desenrolar das descobertas da criança: um estudo sobre as microgêneses cognitivas (Le cheminement des découvertes de l'enfant: recherche sur les microgenèses cognitives). Artes Médicas, Porto Alegre Brazil
13. Jaques PA. et al (2004) Applying affective tactics for a better learning. Proceedings ECAI. IOS Press, Amsterdam
14. Jensen FV, Olsen KG, Andersen SK (1990) An algebra of Bayesian belief universes for knowledge-based systems. Networks, John Wiley, New York 20:637-659

15. Jonassen D (1994) Thinking technology. Educational Technology 34:34-37
16. Mayo M, Mitrovic A (2001) Optimizing ITS behavior with Bayesian networks and decision theory. IJAIED 12:293-309
17. Mullier D, Moore DJ (1998) A web based intelligent tutoring system. Proceedings NETIES'98. Leeds Metropolitan University, Leeds 5-9
18. Murphy E (1997) Constructivist checklist. In: Constructivism: from philosophy to practice. <http://www.stemnet.nf.ca/˜elmurphy/emurphy/cle5a.html> Access on may 2005
19. Murray C, VanLehn K (2000) DT tutor: a decision-theoretic, dynamic approach for optimal selection of tutorial actions. Intelligent Tutoring Systems: 5th International Conference (ITS 2000). Springer, New York 153-162
20. Pearl J (1993) Belief networks revisited. Artificial Intelligence. Elsevier 59:49-56
21. Piaget J (1995) A equilibração das estruturas cognitivas: problema geral do desenvolvimento (L'équilibration des structures cognitives: problème central du développment). Zahar, Rio de Janeiro
22. Piaget J (1978a) Fazer e compreender (Réussir et comprendre). Melhoramentos/EDUSP, São Paulo
23. Piaget J (1978b) A tomada de consciência (La prise de conscience). Melhoramentos/EDUSP, São Paulo
24. Piaget J (1983) A epistemologia genética, sabedoria e ilusões da filosofia, problemas de psicologia genética (L'Épistémologie génétique, sagesse et illusions de la philosophie, problèmes de psychologie gènétique). Abril Cultural, São Paulo
25. Piaget J (1985) O possível e o necessário: evolução dos possíveis na criança (Le possible et le nécessaire: l'évolution des possibles chez l'enfant). Artes Médicas, Porto Alegre Brazil v1
26. Piaget J (1986) O possível e o necessário: evolução dos necessários na criança (Le possible et le nécessaire: l'évolution du nécessaire chez l'enfant), Artes Médicas, Porto Alegre Brazil v2
27. Roesler M, Hawkins D (1994) Intelligent agents: software servants for an electronic information world (and more!). Online 19:19-32
28. Russel S Norvig P (1995) Artificial intelligence: a modern approach. Prentice-Hall, New Jersey 1995
29. Seixas L, Flores CD, Viccari RM, Ladeira M (2002) An architecture for an intelligent learning environment with a constructivist approach. Workshop Proceedings ITS 2002. San Sebastian, 1:8-15
30. Seixas L, Flores CD, Gluz J, Vicari RM (2004) Acompanhamento do processo de construção do conhecimento por meio de um agente probabilístico. Anais 15th SBIE UFAM, Manaus Brazil
31. Seixas LJ (2005) Estratégias pedagógicas para um Ambiente Multiagente Probabilístico Inteligente de Aprendizagem – AMPLIA. Phd Thesis, CINTED/UFRGS, Porto Alegre Brazil
32. Self J (1999) The defining characteristics of intelligent tutoring systems research: ITSs care, precisely. IJAIED 10:350-364
33. Vicari RM, Flores CD, Silvestre AM, Seixas LJ, Ladeira M, Coelho H (2003) A multi-agent intelligent environment for medical knowledge. Artificial Intelligence in Medicine. Elsevier, 27:335-66

5

Tracing CSCL Processes

Cesar A. Collazos[1], Manuel Ortega[2], Crescencio Bravo[3], and Miguel A. Redondo[4]

[1] Department of Systems, University of Cauca, FIET, Sector Tulcan, COLOMBIA ccollazo@unicauca.edu.co
[2] Escuela Superior de Informtica, University of Castilla - La Mancha, Paseo Universidad 4, 13071 Ciudad Real, SPAIN Manuel.Ortega@uclm.es
[3] Escuela Superior de Informtica, University of Castilla - La Mancha, Paseo Universidad 4, 13071 Ciudad Real, SPAIN Crescencio.Bravo@uclm.es
[4] Escuela Superior de Informtica, University of Castilla - La Mancha, Paseo Universidad 4, 13071 Ciudad Real, SPAIN Miguel.Redondo@uclm.es

Group processing and performance analysis exists when groups discuss their progress and decide which behaviors to continue or change. This chapter presents the experience we have developed using a software tool called TeamQuest that includes activities that provide the opportunity for students to examine the performed task from different perspectives, needed to enable learners to make choices and reflect on their learning both individually and socially. We include a model that intend to evaluate the collaborative process in order to improve it based on the permanent evaluation and analysis of different alternatives. Our experience is based on tracing all the activities performed during a computer-supported collaborative activity similar to the affordances of the video artifact, through pauses, stops, and seeks in the video stream. Finally, we discuss how the tracing CSCL process could be used in situations where an analysis based on the expert knowledge of the domain and on the user behavior guides the user to reach the best possible solution using DomoSim-TPC environment.

5.1 Introduction

The management of large classrooms is a serious problem in many programs. Many researchers have shown that working alone could be not beneficial in order to acquire a real knowledge. In the Computer Science field, for example, lectures alone, without student interaction in small groups and laboratories, are an inadequate learning mechanism for Computer Science Education [29]. Collaborative learning has been proven to have many advantages in the

Cesar A. Collazos et al.: *Tracing CSCL Processes*, Studies in Computational Intelligence (SCI) **44**, 103–116 (2007)
www.springerlink.com

teaching-learning processes. It has been agreed upon that before being stated effective, collaborative learning must follow certain guidelines and must have certain roles defined [5]. However, the definition of these guidelines and roles will not guarantee that learning will be achieved in the most efficient manner. It is necessary to define an outline of collaboration where the instructor knows when and how to intervene in order to improve that process. As Katz mentioned, one of the main problems that the teacher must solve in a collaborative environment consists of identifying when to intervene and of knowing what to say [18]. It is necessary for the teacher not only to monitor the activities of a particular student but also the activities of their peers to encourage some kind of interaction that could influence the individual learning and the development of collaborative skills, such as giving and offering help and receiving feedback, agreeing and disagreeing, and identifying and solving conflicts [9, 16, 31].

Just as important as how to evaluate, it is important to mention that how and when to intervene are aspects that could be difficult to realize in an efficient manner if they are managed in a manual way, specially when taking into account that the facilitator could be collaborating with other groups of apprentices in the same class at the same time [7].

The use of computer tools allows the simulation of situations that would otherwise be impossible in the real world. As Ferderber mentioned, the supervision of humans cannot avoid being subjective when observing and measuring the performance of a person [11]. That is why the monitoring carried out by computer tools can give more accurate data as done by people in an manual way.

Monitoring implies reviewing success criteria, such as the involvement of the group members in reviewing boundaries, guidelines and roles during the group activity. As Verdejo mentions, it is useful to interactively monitor the learners while they are solving problems [30]. It may include summarizing the outcome of the last task, assigning action items to members of the group, and noting times for expected completion of assignments. The beginning and ending of any group collaboration involve transition tasks such as assigning roles, requesting changes to an agenda, and locating missing meeting participants [6]. Group processing and performance analysis exists when groups discuss their progress and decide which behaviors to continue or change [15]. So, it is necessary that participants evaluate the previous results obtained in order to continue, by evaluating individual and group activities, and according to the results defining a new strategy to solve a problematic situation. It is also necessary for members of the group to take turns when questioning, clarifying and rewarding their peers' comments to ensure their own understanding of the team interpretation of the problem and the proposed solutions. "In periods of successful collaborative activity, students' conversational turns build upon each other and the content contribute to the joint problem solving activity" [31]. In this paper we present a mechanism based on tracing the collaborative process, a similar experience to the analysis presented in a video protocol analysis [22]. We present a experience using a software tool showing the importance of examine the performed task from different perspectives, needed to enable learners to make choices and reflect on their learning both individually and socially. Section 5.2 presents the related work; our model is described in Section 5.3. Section 5.4 discusses an experience we have developed using a software tool called TeamQuest. Section 5.5 discusses possible scenarios where we can experiment tracing CSCL process and finally the summary is presented.

5.2 Related work

Tutored Video Instruction (TVI) is used to enhance distance learning with interactive multimedia and small group discussion in a collaborative learning methodology in which a small group of students studies a videotape of a lecture. First developed by James Gibbons in the 1970's, TVI uses videotaped lectures as the primary source of instruction for small groups of students outside the traditional lecture classroom. A tutor is present in order to facilitate discussion of the content, thus making TVI a hybrid instructional method intended to "combine the positive features of lectures with those of small group discussions" [12].

Tutor-facilitated classes take advantage of the affordances of the video artifact, through pauses, stops, and searches in the video stream. Students and tutors can use this flexibility of video-based technologies to ask clarifying questions, make challenges to the presentation, review unclear lecture segments, or interject other relevant data, such as anecdotal evidence from personal experiences. Research by Gibbons and others [12, 20] has shown that students learning in this mode achieve at an equal or higher level than their on-campus lecture counterparts.

Gibbons lists several factors he believes are crucial to the success of TVI:

- Attitude, personality, instructional style, and experience of the tutor
- Group size of 3-8 students is optimal for promoting tape stops and discussion
- Student commitment to degree program (or similar educational objective)
- Active classroom participation of students in the lecture class
- Organization of the lecturer, including knowledge of subject and physical presence
- Reduction of outside pressures, such as scheduling and social conflicts
- Ongoing evaluation as managed by a designated administrator

CEVA: Tool for Collaborative Video Analysis is a prototype synchronous Collaborative Video Analysis tool [4]. It provides equivalent functionality to many single-user video analysis tools, CEVA supports several powerful visualisation techniques to support browsing and searching of the logged data. The system enables both synchronous and asynchronous collaboration, synchronous multithreaded event logging, an animated direct manipulation interface, symbolic notation and visualization at different levels, quantitative analysis such as event counts and duration, event search and reordering of video segments. The groupware user interface to CEVA is almost identical to the single-user interface. The only additional interface elements are telepointers which show the location of each user's cursor. Communicating each user's location of activity is a fundamental requirement of WYSIWIS (What You See Is What I See) groupware [28, 13]. Telepointers in all parts of CEVA support gestures and deixis by the multiple users.

Our approach uses similar ideas to TVI and CEVA work, where groups can discuss in a collaborative manner what they have done through work analysis. Next section presents our model that intends to evaluate the collaborative process with the final intention of improving it.

5.3 Our model

Fig. 5.1 represents the model which we intend to develop based on a constant evaluation given that it has to be a dynamic, open, contextualized process that is carried out for a long period of time; it is not an isolated action. Several successive steps of the mentioned process must be carried out in order to obtain the three essential and indispensable characteristics of all evaluations:

- Obtain information: apply valid and reliable procedures to obtain systematic, rigorous, relevant and appropriate data and information that will lay down the foundation of consistency and reliability of the results of the evaluation.
- Make judgments: the obtained data must assure the analysis and the assurance of the facts that we intend to evaluate, so the most appropriate judgment can be obtained.
- Decision-making: in accordance with the emitted assurance of the available relevant information, we can make decisions that come allow with every case.

As it has been mentioned, our model intends to evaluate the collaborative process with the final intention of improving it. That is why we shape the process in such a way that it can be evaluated with the intention of obtaining information that will allow to make judgments of character, with respect to the quality of the collaborative process of the analyzed groups, and as a result it will allow us to make decisions with the intentions of being able to improve the weaknesses observed in the groups. As mentioned by Castillo [3], the main goal of the evaluation is the educative process of the students because this is the end product of the teaching-learning process. With the model we intend to use in this investigation it is possible to supply the teacher with enough elements of judgment to allow him to make decisions in a more confident manner. The mentioned evaluation must be guided, regulated and motivated in order to be conducted through the whole process in such a way that it will cause an improvement in the educative process.

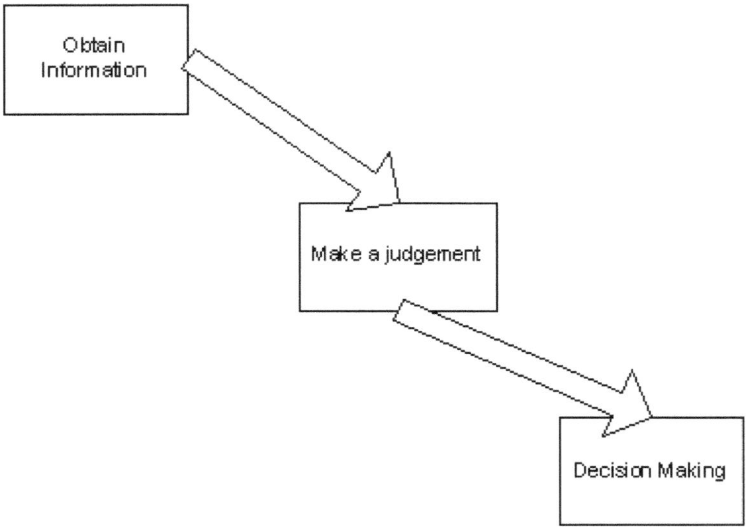

Fig. 5.1. Evaluation Process

5.4 Our experience

Collaborative learning is a complex phenomenon, and studies are being conducted from many different analytical levels and from a range of various theoretical and methodological perspectives. Understanding group dynamics and the collaborative processes of decision making and learning in groups is important both for learners and instructors in collaborative learning programs. Understanding and analyzing the collaborative learning process requires a fine-grained sequential analysis of the group interaction in the context of learning goals. We may notice that supporting individual learning requires an understanding of individual thought process, whereas supporting group learning requires an understanding of the process of collaborative learning [27].

Several researchers in the area of cooperative work take as a success criterion the quality of the group outcome. Nevertheless, recent findings are giving more importance to the quality of the "cooperation process" itself. Success in collaborative learning subject matter means both learning the subject matter (collaborating to learn), and learning how to effectively manage the interaction (learning to collaborate). The knowledge acquisition process for systems supporting collaborative learning warrants a closer look in light of this additional complexity [27]. Traditional instruction tends to emphasize the product of the design and development process, but not the process itself [21].

In order to achieve a good performance in the collaboration process, our experiences has included a group analysis of activities performed during certain time that permit to define new strategies to solve the problematic situation.

All actions performed by the group can be reconstructed by means of logs analysis. All information is recorded by the applications we have developed in order to analyze the mistakes of the group based on a WWSIWWD (What We See Is What We Did) schema. Next we describe the software tool we have developed.

5.4.1 The tool

TeamQuest is a collaborative web game which is played by four persons, each one with a computer. The computers are physically distant and the only communication allowed is computer-mediated. All activities by participants are recorded for analysis and players are made aware of that. The game goal is to go from an initial to a final position through a labyrinth, with the highest possible score, avoiding obstacles and picking the necessary items to carry out the mission in the way. The time spent in the trip is also considered only in case of a tie [8].

Each member of the work group can see only a portion of the game scenario. The members' information needs to be shared if the group wants to achieve its goal. That aspect corresponds to the positive resource interdependence, which relies on the fact that each individual owns specific resources needed for the group as a whole to succeed [17]. The difficulty level of the game - which can be adjusted - is relatively high. Therefore, the group must define and apply a good strategy in order to solve the labyrinth.

The participants are given very few details about the game before playing, and they must discover most of the rules while playing. They also have to develop joint strategies to succeed. The players of a team must reach a common goal satisfying sub-goals in each of the four game stages. Each player is identified with an avatar and name (Fig. 5.2).

The TeamQuest main user interface has three well-defined areas: labyrinth, communication and information. The labyrinth area has four quadrants, each one assigned to a player who has the "doer" role (active player), while the other three players are collaborators for that quadrant. In a quadrant, the doer must move the avatar from the initial position to the "cave" that allows pass to the next quadrant. In the way, the doer must circumvent all obstacles and traps of the map (these obstacles are not visible to all players). Moreover, the doer must pick items useful to reach the final destination. The user interface has many elements showing awareness: the doer's icon, score bars, items which were picked up in each quadrant, etc. (Fig. 5.2).

The communication zone is located at the right hand side of the main screen and it has several windows with faces characterizing each participant avatar. Each participant has a window to write text messages, a receiver selector, and a send button. Also, there are three other windows, similar to the message writing window, which display the messages received from the other players. The information zone displays information about the game status, obstacles, traps, individual views of the game, and game final results.

Fig. 5.2. TeamQuest user interface

The team game score is computed based on the individual score of each player, shown in the score bars. These individual scores start with a predefined value and they are reduced or increased whenever a player's avatar collides with a trap or gets a reward (life potion). The group final score is the addition of the individual scores.

Window video analysis

The Window Video Analysis is a discussion environment. Group members can use it during a break. Breaks may be done at any time during the play. They provide opportunities for analysis of the work done, thus allowing the definition and reinforcement of the common goals. Establishing common goals is part of constructing common grounds, since actions cannot be interpreted without referring to the shared goals, and reciprocally, goal discrepancies are often revealed through disagreements on action [10]. Members of a group do not only develop shared goals by negotiating them, but they also become mutually aware of these goals.

In this activity group member can observe what they have been done, tracing step by step all movements and messages sent/received by the group. In that way, they can observe where they made a mistake, where they hit over an obstacle, etc. This information will permit group members analyze in a simultaneous way the process done in a certain time generating an interesting discussion environment. The log of events on any graphical thread can be changed by rewinding to the appropriate point, clicking the thread's pencil to turn logging on, restarting the tape, and logging over the old thread record.

Textual transcriptions can be edited with normal text-editing procedures in a similar way as CEVA [4].

The application has a structured chat-style user interface, through which the group conversation is held. The application records every message sent by any member of the group. Along with each message, it records the time of occurrence, sender, addressee and current quadrant (the mouse location -X and Y position- when the message was sent). The table 5.1 shows an example of the information gathered by the application. In addition, it records the partial scores and total score by quadrant. The tool also registers the start and finish time of the game, the time spent in each quadrant, and the number of times each player looked at the partial and total scores by quadrant.

Table 5.1. Log file content

X	Y	Quad	From	To	Message	Time
1	1	1				12:00:30
			Claudia	Hans	I need your coordinates	12:00:41
			Claudia	Arxel	I need your coordinates	12:00:52
			Claudia	Lancer	I need your coordinates	12:01:13
			Arxel	Claudia	A2 and F4	12:01:25
			Hans	Claudia	A5 and G5	12:02:08
			Claudia	Hans	D3 and g3	12:03:13
			Lancer	Claudia	Ok	12:03:21
1	2	1			12:03:23	
			Arxel	Claudia	Letters are arrows	12:04:32
1	3	1			12:04:40	
2	3	1			12:04:42	
2	4	1			12:04:45	
2	5	1			12:04:48	
3	5	1			12:04:51	

In this way it is possible to reconstruct all the process done during a collaborative activity, due to all information is ordered according to the execution time and the learner can playback it any time.

According to the table 5.1, we can observe activity begins at 12:00:30 and the initial position is 1,1 of first quadrant. At 12:00:41 Claudia sends a message to Hans and so on. Every time that a new movement occurs, a new position is presented to the group members. Fig. 5.3 depicts the interface of the Window Video Analysis. The user, who begins a new window video analysis, will be the main user who can use the control panel in order to manage the showing velocity of the movements and messages (4). This user can decide to go to the beginning of the activity or to a certain time. In (6) group members can observe all movements performed by the group and (5) shows the messages the group have received/sent during certain time. In a similar way, group members can discuss what they are watching and therefore a chat is presented, where they can write messages (2), send them (3) and watch all the messages (1).

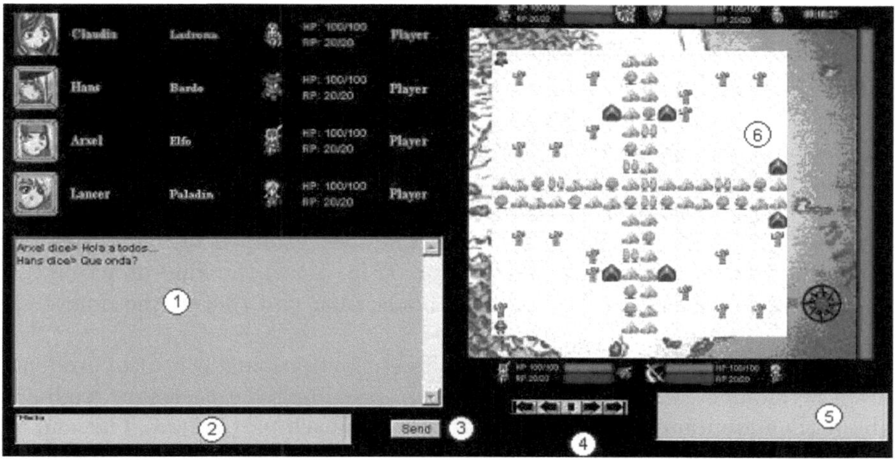

Fig. 5.3. Window Video Analysis Interface

5.5 Discussion

One of the applications of the schema proposed before could be in situations where an analysis based on the expert knowledge of the domain and on the user behaviour guides the user to reach the best possible solution as in DomoSim-TPC environment that is a distributed environment with support for distance learning of domotical design [1]. With this software tool the teaching of domotical design subject in Secondary and Higher Education involves a modification of the educational protocol, while continuing to be based on the paradigm of the PBL (Problem Based Learning) [26] and its application to collaborative environments [19].

First, the teacher carries out a presentation of basic theoretical contents. Next, students are organized in small groups to whom the teacher assigns the resolution of design problems selected from an available library of problems [24]. The characteristics of a problem are described by attributes, including specific didactic objectives. A problem can be selected according to the teacher's approach. This information is used to induce the adaptation of a general resolution schema for a kind of problem. This adaptation is carried out according to the needs, restrictions and help level established by the teacher. The students use PlanEdit [23] to plan the design strategy in order to build a model they consider will satisfy the requirements of the proposed problem.

This plan will be refined in a model at a later stage. Later on they can discuss their proposals. The system offers facilities to collaboratively comment and justify the design decisions taken. Additionally, they can carry out a simulation of a model to check its behaviour under extreme circumstances; and by so doing; they can test if their solution fulfils the requirements [2].

Fig. 5.4 shows the user interface of PlanEdit [25]. This presents different areas: the problem formulation, the list of tasks to carry out, the icon bars representing design actions/operators, the sequence of design actions already planned the current action under construction and a set of buttons used to support several general functions.

The design actions that the student can choose are displayed in the user interface by means of icons in toolbars. They are grouped in four categories according to the components of an action: (Fig. 3-4.a) the kind of action, (3-4.b) the management area, (3-4.c) the house plan and (3-4.d) the domotical operator.

In PlanEdit we can observe the different actions made by the students in every moment of the resolution process, providing the necessary feedback about the individual and collective proposals of each participant. This can be done in the asynchronous tool of Domosim-TPC easily viewing the actions tree and going to the specific branch that we need to observe. We are currently implementing the window video analysis of the actions made in the synchronous tool of Domosim-TPC, permitting in the future to observe the actions performed at any time, according to the approach we are presenting.

In such scenarios our proposed model could be useful because it will give discussion spaces in order to analyze the mistakes that the group have performed and try to find out solutions to another problematic situation. The participants could use this option to stop the simulation process, to propose a question, to cause a reflection, etc., and to continue with the simulation afterwards. This, together with the possibility of the teacher taking part in the simulation with any simulation action, offers interesting ways of mediating in the students' learning [2]. Team potential is maximized when all group members participate in discussions. Building involvement for group discussions increases the amount of information available to the group, enhancing group decision making and improving the participants' quality of thought during the work process [14]. Therefore, encouraging active participation could increase the likelihood that all group members understand the strategy to solve a problematic situation, and decreases the chance that only a few participants understand it, leaving the others behind.

5.6 Summary

We have presented a model that includes activities that provide the opportunity for students to examine the task from different perspectives, needed to

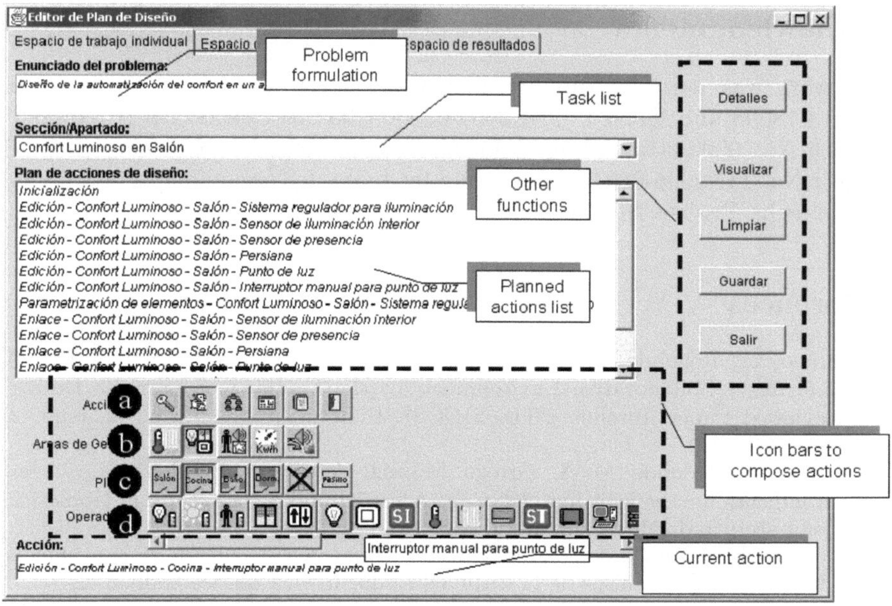

Fig. 5.4. PlanEdit user interface in the Domosim-TPC environment

enable learners to make choices and reflect on their learning both individually and socially.

The model proposed is based on tracing all the activities performed during a collaborative activity similar to the affordances of the video artifact, through pauses, stops, and seeks in the video stream. Therefore, it is much more useful for the students to be able to see the way they constructed a solution to a problematic situation, rather than analyzing only the final result, because our schema includes record and replay mechanisms that permit recovering and reconstructing collaboration processes in a shared scenario. The option we have developed in TeamQuest allows a learner to reflect on the social thinking processes during a collaborative activity and thus colectivelly reexamine, through such reflection, the understanding of those involved.

We have presented a scenario where our model can be used in an appropriate way using DomoSim-TPC. As further work we are going to experiment with some students in order to analyze some usability problems and whether the model proposed supports knowledge in an appropriate manner.

Acknowledgments

This work was partially supported by Colciencias (Colombia) Project No. 4128-14-18008 and CICYT TEN2004-08000-C03-03, Colciencias (Colombia) Project No. 030-2005 and by Ministerio de Educacin y Ciencias (Espaa) Project No. TIN2005-08945-C06-04 and by Junta de Comunidades de Castilla - La Mancha Project No. PBI-05-006.

References

1. Bravo, C., Redondo, M.A., Bravo, J., and Ortega, M., DOMOSIM-COL: A Simulation Collaborative Environment for the Learning of Domotic Design. Reviewed Paper. Inroads - The SIGCSE Bulletin of ACM, vol. 32, num. 2, pp.65-67, 2000.
2. Bravo, C., Redondo, M. A., Ortega, M., and Verdejo, M.F., Collaborative environments for the learning of design: A model and a case study in Domotics. Computers and Education, 46 (2), pp. 152-173, 2006.
3. Castillo, S., Didctica de la evaluacin. Hacia una nueva cultura de la evaluacin educativa. Compromisos de la Evaluacin Educativa, Prentice-Hall, 2002.
4. Cockburn, A., and Dale, T., CEVA: A Tool for Collaborative Video Analysis. In: Payne, Stephen C., Prinz, Wolfgang (ed.): Proceedings of the International ACM SIGGROUP Conference on Supporting Group Work 1997. November 11-19, 1997, Phoenix, Arizona, USA. p.47-55, 1997.
5. Collazos, C., Guerrero, L., and Vergara, A., Aprendizaje Colaborativo: un cambio en el rol del profesor. Memorias del III Congreso de Educacin Superior en Computacin, Jornadas Chilenas de la Computacin, Punta Arenas, Chile, 2001.
6. Collazos, C., Guerrero, L., Pino, J., and Ochoa, S. Evaluating Collaborative Learning Processes. Proceedings of the 8th International Workshop on Groupware (CRIWG'2002), Springer Verlag LNCS, 2440, Heidelberg, Germany, September, 2002.
7. Collazos, C., Guerrero, L., and Pino, J. Computational Design Principles to Support the Monitoring of Collaborative Learning Processes; Journal of Advanced Technology for Learning, Vol.1, No. 3, pp.174-180, 2004
8. Collazos, C., Guerrero, L.,Pino, J., and Ochoa, S., A Method for Evaluating Computer-Supported Collaborative Learning Processes; International Journal of Computer Applications in Technology, Vol. 19, Nos. 3/4, pp.151-161, 2004
9. Dillenbourg, P., Baker, M., Blake, A. and O'Malley, C., The evolution of research on collaborative learning. In Spada, H. and Reimann, P. (eds.), Learning in Humans and Machines: Towards an interdisciplinary learning science. pp-189-211. Oxford: Elsevier, 1995.
10. Dillenbourg, P., What do you mean by collaborative learning?. In P. Dillenbourg (Ed). Collaborative Learning: Cognitive and Computational Approaches. Pp. 1-19, Oxford:Elsevier, 1999.
11. Ferderber, C., Measuring quality and productivity in a service environment. Indus. Eng. Vol. 13, No. 7, pp. 38-47, 1981.
12. Gibbons, J. F., Kincheloe, W. R., and Down, K. S., Tutored videotape instruction: A new use of electronics media in education. Science, 195, 1139-1146, 1977.

13. Gutwin, C., Stark, G., and Greenberg, S., Support for workspace awareness in educational groupware. In ACM Conference on Computer Supported Cooperative Learning (CSCL '95). Bloomington, Indiana. October 17-20, 1995, pp. 147-156. Lawrence ErlbaumAssociates, Inc, October 1995.
14. Jarboe, S., Procedures for enhancing group decision making. In B. Hirokawa and M. Poole (Eds.), Communication and Group Decision Making, pp. 345-383, Thousand Oaks, CA:Sage Publications, 1996.
15. Johnson, D., Johnson, R., and Holubec, E., Circles of learning: Cooperation in the classroom (3rd ed.), Edina, MN: Interaction Book Company, 1990.
16. Johnson, D., Johnson, E., and Smith, K., Increasing College Faculty Instructional Productivity, ASHE-ERIC Higher Education Report No.4, School of Education and Human Development, George Washington University, 1991.
17. Johnson, D., Johnson, R., and Holubec, E., Cooperation in the classroom, 7th edition, 1998.
18. Katz, S., and O'Donell, G., The cognitive skill of coaching collaboration. In C. Hoadley & J. Roschelle (Eds.), Proceedings of Computer Supported for Collaborative Learning (CSCL), pp. 291-299, Stanford, CA., 1999.
19. Koschmann, T, Kelson, A.C., Feltovich, P.J. and Barrows, H., Computer-Supported Problem-Based Learning: A Principled Approach to the Use of Computers in Collaborative Learning. Koschmann, T. (Ed.) CSCL: Theory and practice of an emerging paradigm. Lawrence Erlbaum Associates, 1996.
20. Lewis, J. L. and Blanksby, V., New look video in vocational education: What factors contribute to its success? Australian Journal of Educational Technology, 4 (2), 109-117, 1998.
21. Linn, M.C., and Clancy, M.J., The Case for Case Studies of Programming Problems. Communications of the ACM, Vol. 35,No. 3, pp. 121-132, 1992.
22. Neal, L., The use of video in empirical research. ACM SIGCHI Bulletin: Special Edition on Video as a Research and Design Tool, 21(2):100-101, 1989.
23. Redondo, M. A., Bravo, C., Ortega, M., and Verdejo, M. F., PlanEdit: An adaptive tool for design learning by problem solving. In P. De Bra, P. Brusilovsky, & R. Conejo (Eds.), Adaptive hypermedia and adaptive web-based 670 systems, LNCS (pp. 560-563). Berlin: Springer, 2002.
24. Redondo, M.A., Bravo, C., Bravo, J., and Ortega, M.,. Applying Fuzzy Logic to Analyze Collaborative Learning Experiences. Journal of the United States Distance Learning Association. Special issue on Evaluation Techniques and Learning Management Systems. Vol. 17, N 2, 19-28, 2003.
25. Redondo, M.., Bravo, C., Ortega, M., Verdejo, M.F.; Providing adaptation and guidance for design learning by problem solving. The Design Planning approach in DomoSim-TPC environment, Computers & Education. An International Journal. To be published, 2006
26. Savery, J., and Duffy, T., Problem based learning: An instructional model and its constructivist framework. In B. Wilson (Ed.), Constructivist learning environments: Case studies in instructional design. Englewwod Cliffs, NJ: Educational Technology Publications. pp. 135-148, 1996.
27. Soller, A., and Lesgold, A., Knowledge Acquisition for Adaptive Collaborative Learning Environments. AAAI Fall Symposium: Learning How to Do Things, Cape Cod, MA, 2000.
28. Tatar, D., Foster, G., and Bobrow. DG., Design for conversation: Lessons from cognoter. International Journal of Man-Machine Studies, 34(2):185-209, 1991.

29. Tucker, A., and Wegner, P., Computer Science and Engineering: the Discipline and Its Impact. CRC Handbook of Computer Science and Engineering. CRC Press, Boca Raton, December 1996
30. Verdejo, M.F.,. A Framework for Instructional Planning and Discourse Modeling in Intelligent Tutoring Systems. In E. Costa (ed.), New Directions for Intelligent Tutoring Systems. Springer Verlag: Berlin, pp. 147-170, 1992.
31. Webb, N., Testing a theoretical model of student interaction and learning in small groups. In R. Hertz-Lazarowitz and N. Miller (Eds.), Interaction in cooperative groups: The theoretical anatomy of group learning, pp. 102-119. NY: Cambridge University Press, 1992.

6

Formal Aspects of Pedagogical Negotiation in AMPLIA System

João Carlos Gluz[1], Cecilia Dias Flores[2], and Rosa Maria Vicari[3]

[1] Post Graduation Program in Applied Computer Science – PIPCA, Universidade do Vale do Rio dos Sinos – UNISINOS, 93.022-000, São Leopoldo, RS, Brasil jcgluz@unisinos.br

[2] Informatics Institute, Federal University of Rio Grande do Sul – UFRGS, POBox:15064, 91501-970,Porto Alegre, RS, Brazil dflores@inf.ufrgs.br

[3] Informatics Institute, Federal University of Rio Grande do Sul – UFRGS, POBox:15064, 91501-970,Porto Alegre, RS, Brazil rosa@inf.ufrgs.br

This chapter presents a pedagogical negotiation model developed for AMPLIA, an Intelligent Probabilistic Multi-agent Learning Environment. It focuses on the formal aspects of the negotiation process, trying to abstract the most general characteristics of this process. The negotiation is characterized by: i) the negotiation object (belief on a knowledge domain), ii) the negotiation initial state (absence of an agreement, which is characterized by an unbalance between credibility, confidence, and a low BN model quality); iii) the final state (highest level of balance between credibility and confidence, and good BN model quality); and iv) the negotiation processes (from state ii to state iii). Three intelligent software agents: Domain Agent, Learner Agent and Mediator Agent were developed using Bayesian Networks and Influence Diagrams. The goal of the negotiation model is to increase, as much as possible: (a) the performance of the model the students build; (b) the confidence that teachers and tutors have in the students' ability to diagnose cases; and the students' confidence on their own ability to diagnose cases; and (c) the students' confidence on their own ability to diagnose diseases. The challenge was to create a learning environment that could really use the key concepts embedded in the idea of negotiation in a teaching-learning process (pedagogic negotiation), aiming at setting the project's principles, which are: Symmetry between man and machine and existence of negotiation spaces.

João Carlos Gluz et al.: *Formal Aspects of Pedagogical Negotiation in AMPLIA System*, Studies in Computational Intelligence (SCI) **44**, 117–146 (2007)
www.springerlink.com

6.1 Introduction

This chapter discusses the pedagogical negotiation process (PN) involved in the implementation of a real learning environment – AMPLIA [25, 8, 9]. It focuses on the formal aspects of the negotiation process, trying to abstract the most general characteristics of this process. Pedagogical negotiation is a complex process that tries to bring to Intelligent Learning Environments (ILE) research area the key concepts embedded in the idea of negotiation in a teaching-learning process. AMPLIA aims to incorporate this negotiation process directly in its system's design principles, intending to achieve: symmetry between man and machine and existence of negotiation spaces between users and the system. The challenge is in the search of symmetry between man and machine. Such symmetry provides the same possibilities to the user and the system as to actions, and symmetric rights for decision taking. In an asymmetry mode, an agent has always the decisive word; there is no space for a real negotiation. In the symmetric mode, there is no pre-defined winner; conflicts need to be solved through a negotiation. Precisely, the cognitive processes that trigger an explicit declaration, justify an argument or refuse the partner's point of view are most likely to explain the reasons why collaborative learning is more efficient than lonely learning.

The main functions of ILE (explanation, education, diagnostic) are traditionally implemented as one-way mechanisms; this means the system has the total control. AMPLIA, however, tries to treat them as bilateral processes, this means that a diagnostic model is not built by the student and sent to the system. This model is built collaboratively, and there are some negotiation moments. It is clear that for a negotiation to take place there must be a level of latitude available to agents, otherwise anything can be negotiated. This defines the global negotiation space within which the two agents try to build a shared understanding of the problem and its solution. The negotiation naturally does not take place in a single plan. Dillenbourg et al. [5] say that human partners do not negotiate a single shared representation, but they really develop several shared representations, that is, they move in a mosaic of different negotiation spaces.

The application of formal approaches to understand or conceptualise aspects of educational processes, including negotiation aspects, is a relatively new research area, at least in the sense of the ILE or ITS (Intelligent Tutoring Systems) formalisation. Self and Dillenbourg [24, 22, 4] present important works in this area. The paper starts the considerations from the analysis of the formal framework proposed by these authors, and show the questions and problems found in the application of these formal techniques to model the communication of AMPLIA system. A new formal model is proposed to solve these problems, and the interactions that occur in the AMPLIA are shown to be particular cases of negotiations, classified as Pedagogical Negotiations. An abstract view of this kind of negotiation is presented, following the approach suggested by Jennings, Faratin and others [16] to automatic negotiation.

The chapter is organised as follows: in Sect. 6.2 the literature is reviewed and the most important questions are discussed. Section 6.3 describes the object of the negotiation that takes place in AMPLIA. Section 6.4 describes the initial and final states of this negotiation. Section 6.5 shows the process involved in negotiation and discusses the roles played by each software agent. Section 6.6 present the theoretical model used in the formal analysis of the PN process and the formalisation aspects of this process. Section 6.7 shows examples and results about the process of pedagogical negotiation. Section 6.8 shows a summary of this chapter.

6.2 Theoretical Basis

The application of formal approaches to understand or conceptualise aspects of educational processes is a relatively new research area, at least in the sense of the formalisation of Intelligent Tutoring Systems (ITS) (or Intelligent Learning Environments – ILE). Self [24, 22] presents solid foundation on the application of formal methods to the analysis of student models. These works show theoretical benefits and problems related to this kind of research. Particularly, in [24] the formal analysis clearly shows that there is a profound relationship between several areas of Artificial Intelligence (AI), like machine learning and cognitive agent modelling, and ITS research. The paper attempts to provide a theoretical and computational basis for student modelling. It also tries to warrant that this formal basis is psychologically neutral and independent of applications. The formal model used in the paper is derived mainly from various areas of theoretical artificial intelligence, particularly from the epistemic/doxastic modal logics (Logic's of Knowledge/Beliefs) and from BDI (Belief, Desire and Intentions) logic's used for agent cognitive and communication modelling. This is the main point of contact between these works of Self and the work presented in this chapter, because the latter will start from a formal model similar to the presented in [24]. However, due to the nature of the knowledge learned in AMPLIA environment (how to design bayesian networks for medical diagnostics purposes) and, most important, due to the probabilistic modelling of the reasoning processes used in the domain and tutoring agents of AMPLIA system, it was necessary to generalise the purely logical BDI model to support probabilities, through the definition of a new probabilistic logical model, that keeps a solid theoretical basis and that is "compatible" with the logical models used as the foundation of current cognitive agent modelling (BDI models) [19, 3, 17].

Another important work in this area is Dillenbourg's [4] paper that describes an abstract formal framework that shows how the basic entities (modules or agents) of an ITS, like the tutor/domain module and the student model module, can be structurally organised in several abstraction layers. Dillenbourg's framework also shows the relationships that occur between the entities of each layer and what kind of knowledge is related to each one of these relationships. The abstract layers of the framework (the 'vertical' dimension) are based upon the computational distinction among behaviour, behavioural knowledge, and conceptual knowledge, while the entity sub-classification used in the framework (the 'horizontal' dimension) is a relatively simple one, assuming only the existence of the "system", "the learner", and "the system's representation of the learner" (the student model). The most important point in this article is that the interaction between the learner and the system was clearly contextualized as a search space problem and methods for establishing the search space for learner models and for carrying out the search process were reviewed. However, this work needs to extend this model, in the sense that in AMPLIA system it is needed to clearly identify the entity responsible by the teaching mediation process and the entity responsible by the learning domain knowledge (the "system" entity must be sub-classified). As the main focus of this work does not lie in the detailed analysis of the student model, but in the properties of the interaction between the student and the entities of the system, it makes a more profound analysis of these kinds of interactions, that the analysis made in [4].

Indeed, in this work it is considered that these kinds of interactions are the most important elementary units for the analysis of teaching and learning process. The present work considers these interactions essentially as negotiations (Pedagogical Negotiations - PN) that occur between the agents involved in this process. Discussions about the use of negotiation mechanisms in learning environments are not recent. According to Self [23], there are two major motivations for the use of negotiation in ITS: i) they make possible to foster discussions about how to proceed, which strategy to follow, which example to look for, in an attempt to decrease the control that is typical of ITS, and ii) they give room for discussions that yield different viewpoints (different beliefs), provided that the agent (tutor) is not infallible.

The approach of PN can be applied to areas of knowledge that share some characteristics such as incomplete knowledge and different points of view or even domains where there is no "knowledge" – considered in its classical definition, in which knowledge is always something true – but a set of justified beliefs about what one can argue and debate. These characteristics foresee the transformation of viewpoints, both from the system and from student, into beliefs instead of knowledge. This implies a special type of teaching dialogue, if an interactive change of justified beliefs is a simplified definition of argumentation [21]. This is a complex process because it involves the student's autonomy, the symmetry of relations between teacher and students or among agents, and the levels of flexibility, which involve the agents' level of freedom to perform their actions [2]. This work does not see the presence of 'conflict' (either openly declared or acknowledged or not) as essential in the definition of negotiation. The basic requirement is that the interaction among agents shares a common goal so that an agreement is reached with respect to the negotiation object. Usually, different dimensions of the negotiation object will be negotiated simultaneously. The initial state for a negotiation to take place is the absence of an agreement, which can include a conflict or not. In the case of a teaching and learning process, a point of conflict is the relation of self-confidence and mutual confidence between teachers and students, besides their own beliefs about the knowledge domain. A process of teaching and learning is a way of reducing the asymmetry between the teacher's and the student's confidence on the studied topic.

These considerations bring this (formal) research directly the questions studied in negotiation protocols for multiagent systems. Following Sandholdm [20] there are six basic criteria that can be used to evaluate these kinds of protocols: (a) Social Welfare; (b) Economic (Pareto) Efficiency; (c) Individual Rationality; (d) Stability; (e) Computational Efficiency; (f) Communication Efficiency. Generally, the (b) criteria can be considered a sub-case of (a) criteria, Pareto efficiency being considered as an economical sound way to achieve social welfare. Criteria (a), (b) and (d) (and eventually (c)) imply in strong correlations between agent's negotiation protocols research and well established negotiation theories from Economical Sciences, like Game Theory and Nash equilibrium[4]. Indeed Sandholm lets it clear that the most general form to handle the stability in a negotiation process is to consider this question like a search for Nash equilibrium. The problem is the profound influence that Utility Theory (and correlated Preference Theory) [26] has on this kind of research. These theories make a good sense in economical terms, but they have not clear application to other kind of negotiations. In particular, there are not clear indications of how they can be interpreted or used in a PN context. On the contrary, we will consider PN as a process of negotiation that occurs in a classroom that only relies on pedagogical strategies to achieve agreements. No economical or monetary means must be used in these agreements.

However, the formal aspects of theories and models used in agent's negotiation protocols research are very valuable and this work tries to apply these formalisation techniques to the problem approached. In particular, this work follows the framework presented by Jennings et al. [16] where the negotiation is modelled as a search space process. The problem of using Utility or Preference Theory is solved here by using a concept that is considered more basic for the educational process: the concept of confidence relationships[5] that can and must be established during the evolution of the educational process. Of course there are several questions related on how to make good assessments or estimations of this confidence and this work should be considered only as a first (formal) approach to this question. However, this work has a positive answer

[4] See [18] and [27] for an introduction to Game Theory applied to AI. Reference [26] presents a broad and general presentation of Utility and Preference Theories and Nash equilibrium applied to micro economical studies.

[5] The notion of confidence adopted here is turned towards an expectation with the future actions of an agent, which is similar to the notion of confidence by Fischer & Ghidini [7]. They base it on a modal logic of beliefs and abilities, which intuitively is according to the idea of considering someone reliable because we know how this person is going to behave in given situations [10].

to this question, at least in AMPLIA domain, based on the idea that these assessments are probabilistic in their nature. Using probabilistic (bayesian) modelling of the reasoning process related to these assessments the authors were able to achieve good practical results.

AMPLIA was designed as an extra resource for the education of medical students [8, 9]. It supports the development of diagnostic reasoning and modelling of diagnostic hypotheses. The learner activities comprise the representation of a clinical case in a Bayesian Network (BN) model (such process is supported by software agents). BN have been widely employed in the modelling of uncertain knowledge domains [14]. Uncertainty is represented by the probability and the basic inference of the probabilistic reasoning, that is, the calculus of the probability of a variable or more, according to the evidence available. This evidence is represented by a set of variables with known values.

The main goal of a PN is to provide and establish a high degree of confidence among the participants of the process. It is not a generic confidence, but a very specific and objective one, associated with abilities that students demonstrate when dealing with learning domain. The degree of belief on an autonomous action is an important component of confidence that will take place in a given teaching and learning process. It will indicate how much students' actions are guided by trials or hypotheses. This variable corresponds to system's credibility on student's actions and its value is inferred by the Learner Agent. Self-confidence (the confidence the student has on the BN model) is another variable used in the pedagogic negotiation, once students are confident on their hypothesis, or at least trust them increasingly, as they build their knowledge. The quality of the BN model is the third element considered in the negotiation process, as the student must be able to formulate a diagnosis that will probable be compliant with the case, as the diagnosis proposed by an expert would be. The Domain Agent evaluates this quality. The Mediator Agent uses these three elements presented above as parameters for the selection of pedagogic strategies and tactics, as well as to define the way in which they will be displayed to the student.

The negotiation is characterised by: i) the negotiation object (belief on a knowledge domain), ii) the negotiation initial state (absence of an agreement, which is characterised by an unbalance between credibility, confidence, and a low BN model quality); iii) the final state (highest level of balance between credibility and confidence, and good BN model quality); and iv) the negotiation processes (from state ii to state iii). This is the base of the negotiation model developed in AMPLIA.

6.3 The Object of Negotiation in AMPLIA

The object of negotiation in AMPLIA is the belief on a diagnostic hypothesis outlined for a clinical case. The student's diagnosis model is built through an editor of BN. From the pedagogic point of view, BN is a tool that students use to represent their knowledge by means of probabilistic models. They can build and observe their study object and formulate and test their hypotheses through those BN model. As AMPLIA is a computational resource, there is an intelligent software agent of the system that aid students build their BN model for the present case.

The teacher's beliefs are also modelled through BN, named expert's model and stored in the environment database. As these beliefs are concerned to an uncertain domain, the expert's BN model can be incomplete, the reason why a base of real cases, which is continuously updated, is used to validate the expert's beliefs. A belief expresses how much someone believes x, or how much x is considered right. Under this viewpoint, the students' and teacher's confidence are two parameters comprised in negotiation. The participants of a negotiation process use the resource of argumentation to persuade the other participants to change their beliefs (or viewpoints). In AMPLIA, the argumentation resources are represented by the teacher's actions, translated into the selection of a strategy that is considered more adequate to convince the student. Students' argumentation is represented by the modifications (or not) of their BN and by the level of confidence they are asked to declare. As the object of the negotiation is a BN model, it is probable that are problems both in its topologic structure and in the tables of conditional probabilities. The problems identified in the student's BN model are classified as in Table 6.1.

Table 6.1. Student's BN model classification

Not feasible	It is not a BN, there are cycles or disconnected nodes
Incorrect	Absence of diagnostic node or diagnostic node represented as father node of symptoms, presence of excluding node with incorrect representation of probabilities
Potential	It has at least one trigger node or one of the essential or additional nodes; presence of unnecessary nodes
Satisfactory	The majority of nodes are essential and additional; some relations are missing
Complete	It has a good performance as compared to the expert's model and to the database of real cases

6.4 The Negotiation Processes

The pedagogical negotiation in AMPLIA can be seen as a way of reducing the initial asymmetry in the confidence relation between teacher/student concerning the topic studied. Such a negotiation is intended to maximise the confidence of all. The following schema shows initial and final expected conditions for the negotiation process:

- Initial state of the pedagogical negotiation process:
  Teacher: (IP.1) High level of confidence in his or her own knowledge.
  (IP.2) Low level of confidence in students' knowledge.
  Student: (IA.1) Low level of confidence in his or her own knowledge.
  (IA.2) High level of confidence in teacher's knowledge.
- Final (expected) state of the pedagogical negotiation process:
  Teacher: (FP.1) High level of confidence in his or her own knowledge.
  (FP.2) High level of confidence in students' knowledge.
  Student: (FA.1) High level of confidence in his or her own knowledge.
  (FA.2) High level of confidence in the teacher's knowledge.

Conditions (IP.1) and (FP.1), as well as (IA.2) and (FA.2) should not change, being only bases for an adequate beginning, development and end of the process. The result of the process would be the increase in the level of confidence that the teacher has on the student: (IP.2) for (FP.2), and of the students on themselves: (IA.1) for (FA.1).

The participants of a pedagogic negotiation are the student and the teacher. In AMPLIA, the student is represented by the Learner Agent and the teacher tasks are performed by three software agents: Learner Agent, Domain Agent and Mediator Agent. Learner Agent, in addition to representing the student, also infers the credibility level, as a teacher that observes the students' actions. Domain Agent evaluates the quality of students' BN model and checks the performance both of students' and teacher's BN models against a database of real cases. Mediator Agent is the agent responsible by the selection of pedagogic strategies and for the successful conclusion of pedagogic negotiation.

Figure 6.1 shows main elements of the negotiation model: initial state, final state, the negotiation object, and negotiation process. Negotiation objects are represented by circles that indicate the status of the student's BN model. Status is labelled as Main Problem, which is identified by the Mediator Agent. The initial state is defined in terms of specific elements, student's and system's individual and mutual goals and beliefs. The only element required is the mutual goal of agreeing on some negotiation object. The final state will be reached when a symmetry between the student's (Self-confidence) and the system's (Credibility) confidence is reached, and when the student's BN model reaches the status Satisfactory or Complete, with a similar or even better performance than the expert's BN model. The negotiation process has the purpose of achieving the final state from the initial state. The inverted triangle in Fig. 6.1 is meant to indicate convergence towards this final state. The pedagogic strategy is selected based on the Main Problem (MP) of the students' BN model and their Self-confidence. Credibility represents the "fine tune" and determines which tactics will be applied to the student, the tactics is meant to be the way in which the strategy will be displayed.

In the initial state, the object of negotiation – students' BN model – is not built yet; therefore, there is no negotiation. The pedagogic strategy used

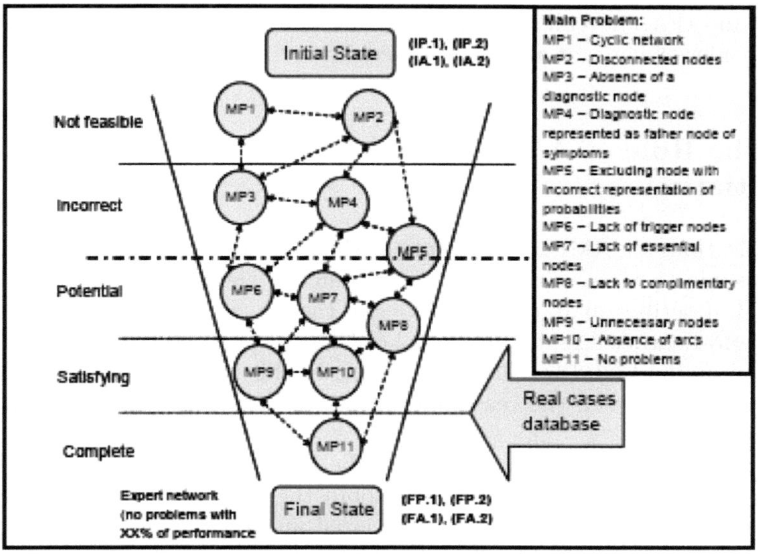

Fig. 6.1. Negotiation process in AMPLIA

in this case will be to guide students: the tactics can be to present a problem or suggest that students check their BN model again and look for conceptual problems. In the following level, in which there is a mistake in the representation of the object, the Mediator Agent disagrees with students' BN model. In these first levels, the focus of the Mediator Agent is on a concrete object (the BN model) and does not include the students' confidence on their BN model. In the following levels, the negotiation process starts: the goal of the Mediator Agent is now to make students reflect and enhance their diagnostic hypothesis represented by the BN model by including lacking nodes and indicating the relationship among them. When BN model created by students starts to enter the satisfactory level (as compared to the expert's model), the Mediator Agent starts to warn the students that some adjustments in the probabilities of the BN model are required. At the same time, students' BN models are submitted to the database of real cases for the evaluation of performance. The expert's BN model is also submitted to this base. The database is continuously updated, so that the Mediator Agent is able to accept BN models built by students that are better than BN models built by experts. It is worth saying that the conditions (FP.1) and (FA.2) are the basis for this process to take place. Even if some student's BN model is classified as complete but the Learner Agent detected low credibility, or if the student declared a low confidence, the Mediator Agent will use different strategies, such as demos or discussions, in order to enhance the model, these actions correspond to the

(FP.2) and (FA.1). While this status is not reached, the Mediator Agent does not consider that negotiation has ended.

6.5 The Role of AMPLIA's Agents in the Pedagogical Negotiation

The role that all AMPLIA's agents perform in PN processes is related to the reasoning processes that occur internally in these agents. Section 4.3 of Chap. 4, describes main characteristics of the reasoning processes of Domain Agent and Learner Agent. Section 4.4 of Chap. 4 presents the teaching tactics selection method used by Mediator Agent. In this section, it is presented the relationships between the reasoning processes and the communication tasks of the agents.

In terms of pedagogical negotiation the Learner agent represents the student, gathering all concrete evidences about the status of its learning process, registering the self-confidence level declared and trying to infer probabilistically – such as a teacher would do – the levels of grasp of consciousness of the student, after the observation of student's actions. This inference will result in the credibility levels (variables) used in AMPLIA. These variables are conditional nodes of the BN and are informed to the Mediator agent.

The Domain agent and Mediator agent share the teacher's role. The Domain agent incorporates the knowledge base on the theme to be studied and, therefore, it has the higher confidence level about the topic. The Mediator agent incorporates negotiation mechanisms needed to solve conflicts of this process. These mechanisms are derived from teaching pedagogical strategies that can be used in pedagogical negotiation.

The negotiation process follows the following protocol:

1. The Domain agent presents a case study to students. The Learner agent only takes notes on the example and passes it to students.
2. The Domain agent made available the case studies from where students model the diagnostic hypothesis. Students model the diagnostic hypothesis, and send (through the Learner agent) their model to the Domain agent to be evaluated. This evaluation refers to the importance of each area in the model (trigger, essential, complementary...).
3. Based on the result of the Domain agent analysis and on the confidence level (declared by the student) supplied by the Learner agent, the Mediator agent chooses the best pedagogic strategy, activating the tactics suitable to a particular situation. In this process, the agent follows the diagram defined in the Fig. 6.1
4. The student evaluates the message received from the Mediator agent and tries to discuss the topics, which considers important, by changing its model. At this stage, the student may also decide to give up the learning process.

The AMPLIA's negotiation process occurs in a dynamic choice of strategies. The parameters considered are linked to student's beliefs, to the evaluation carried out by the Domain agent and to the observations registered by the Learner agent.

In this negotiation process, both the student and the Domain agent have the possibility of giving up the interaction. The Domain agent only leaves the negotiation process when the student presents a solution, whose performance is equivalent or better than its model. The Domain agent may come to accept student's modelling, although it does not correspond exactly to its model, but students use the arguments to solve the study case problem presented to them.

6.6 Formalization of PN Process

The main approach of this work to formalize the PN process was based on the modelling of the communication interactions related to this negotiation process. The reason behind this approach is that a PN process is essentially a communication process were agents exchange bits of knowledge to achieve a common agreement. Thus, to formalize the PN process, it is necessary to model the communication interactions that occur between agents during this process. However, it is possible that a formal model restricted only to the communication processes should not be sufficient, in the sense that the internal reasoning processes of the agents must also be taken into account to fully represent PN processes. To this end, it is needed to be sure that the formal framework being used can also represent the reasoning processes of the agents.

This section will show how the formalization model of PN was built. First, it is shown why current frameworks for agent communication are not enough expressive to represent PN process in AMPLIA system, and then it is presented a new formal framework conceived by the authors that can be used to represent this kind of process. After that, it is shown how PN processes can be represented in this formal framework. First, it is defined the formal structure of the knowledge that is shared between the agents of the system and its related communication space. Then it is shown how the agent interactions that occur in PN processes can be specified as conditions (equations) over the shared knowledge space, defining the negotiation space between these agents. Finally, it is shown how the internal reasoning of the agents can be represented in this formal framework.

6.6.1 Current Agent Communication Frameworks

When the analysis and modelling of the AMPLIA's communication processes was started, it was first tried to apply current standard formalisation techniques and languages used in agent communication research.

It was decided to use only standard languages and protocols to model and implement these communication tasks in order to allow reusability of the AMPLIA's agents knowledge and also to allow an easier interoperation of AMPLIA with others intelligent learning systems. To this purpose it was decided to use FIPA (Foundation for Intelligent Physical Agents) standards [6] based on two assumptions: (a) The standards are a good way to ensure MultiAgent Systems (MAS) knowledge reusability and interoperability. (b) The formal basis of FIPA standards offer an abstract and architecture independent way to model all communication tasks of the system, allowing a high level description of the communication phenomena. The last assumption is a pre-requisite to the formal analysis of interactions and negotiations that occur in AMPLIA system.

The FIPA organisation provides a framework composed of several standard tools that can be used to represent and implement agent's communications tasks. These tools include an Agent Communication Language (ACL) [6], a content language (SL) and several interaction protocols. However, it was found that it was impossible to meet even most basic communication requirements of AMPLIA using only FIPA standard ACL, content languages or interaction protocols. All AMPLIA's agents use probabilistic (bayesian) knowledge and need to exchange this kind of knowledge between them. What it was found is that FIPA standards simply do not have a word to say about the communication of this kind of knowledge. The semantics of FIPA-ACL is entirely based on a purely logical framework; all standard content languages of FIPA have a strong logical "flavour" in its semantics (two of them, KIF and SL are modal logics with a computer friendly syntax). Indeed the logical formal modal that fundament the semantics of FIPA standards assigns no meaning to probabilistic knowledge representation or communication.

Therefore, the authors were motivated to extend the FIPA standards and generalize the theoretical model that fundament these standards to support probabilities. This extension and generalisation was made in two steps: (a) the epistemic modal logic used as the basis for the semantics of FIPA standards was generalised to a probabilistic logic, and (b) new communicative acts were added to FIPA ACL to support the communication of probabilistic knowledge.

6.6.2 The New Formal Framework

The logic used to formalise the negotiation process was based on the logic *SL* (Semantic Logic) that is the modal logic used as the basis for FIPA agent communication standards. This logic was defined by Sadek's works [19]. The extension of the *SL* logic is named *SLP* (*Semantic Language with Probabilities*) [11, 12]. It was defined through the generalisation the *SL* formal model. The basic idea behind that generalisation is to incorporate probabilities in *SL* logic following works from Halpern [13] and Bacchus [1] that integrate probabilities in epistemic (beliefs) modal logics. *SLP* will incorporate, besides numerical operator and relations, terms that denote probabilities expressing the subjective probability (degree of belief) of a given sentence or statement being true.

SLP is a first order probabilistic modal logic with equality. Besides the usual operators and quantifiers of the predicate logic, *SLP* also inherit from *SL* modal operators to express the beliefs ($\boldsymbol{B}(a,\theta)$), choices $\boldsymbol{C}(a,\theta)$ and intentions ($\boldsymbol{I}(a,\theta)$) of an agent a. In this way the belief that some agent a have in proposition P or their choice that some situation Q be true, can be expressed, respectively, by $\boldsymbol{B}(a,\mathrm{P})$ or $\boldsymbol{C}(a,\mathrm{Q})$. It is also possible to build action expressions that can be connected in series $e_1;e_2;...;e_n$, in alternates $e_1|e_2$ or verified by an agent a $(a,e)?$. Temporal and possibility assertions can be made based on the fact that an action or event has happened ($\boldsymbol{Done}(e,\theta)$), on the possibility that an action or event may happen ($\boldsymbol{Feasible}(e,\theta)$), on which agent is responsible for an action ($\boldsymbol{Agent}(a,e,\theta)$), among others. The probabilistic term $\boldsymbol{BP}(a,\theta)$ is specific for *SLP* and informs the probability of a proposition θ be true with respect to the beliefs of agent a, that is, it defines the subjective probability assigned to θ by a. *SLP* also incorporates numerical expressions that can be used to express probabilistic relationships. For example, the formula $\boldsymbol{BP}(a,\exists(x)(\mathrm{P}(x)) \leq \mathbf{1}$ express the fact that the subjective probability assigned by agent a to the possibility that some element of the domain satisfies $\mathrm{P}(x)$ is less than $\mathbf{1}$.

However, one very important restriction in *SLP* is that the θ formula in any $\boldsymbol{BP}(a,\theta)$ term must be a sentence (a closed formula) of the logic. With this restriction and the additional restriction that numerical constants or numerical variables cannot be used as arguments of logical predicates (and vice-versa) it was possible to define an axiomatic system for *SLP* that not only is correct but also does not turn the validity problem of *SLP* formulas undecidable, if it is already not undecidable in *SL* [11].

Like *SL*, *SLP* also can be used as a content representation language for FIPA-ACL communicative acts. This allows the representation and distribution of probabilistic knowledge like bayesian networks (see [11] for details), between agents using standard assertive (**inform**) acts. However, to do this is necessary to assume a particular structure in the contents of these acts. The assertive acts defined in Speech Act theory (and the equivalent **inform** FIPA-ACL acts) do not assume any particular internal structure in the propositions passed as contents of these acts. So, in the general case of probabilistic communication not seems reasonable to always assume a particular structure (bayesian networks, for instance) in the content of assertive act used to communicate probabilities. To handle this it is proposed that the strength (or weakness) of the assertive force of some speech act should be measured by a probability. These new probabilistic communicative acts were considered extensions to the FIPA-ACL language, creating the PACL (*Probabilistic Agent Communication Language*). The acts **inform-bp** and **query-bp** acts were defined, respectively, to allow that the information about subjective probabilities of an agent to be shared with other agents and to allow that a given agent could query the degree of belief of another agent.

6.6.3 Shared Knowledge and Communication Space

To formalize the shared knowledge first is necessary to identify the agents that need to share this knowledge. AMPLIA has three types of artificial entities: Learner, Mediator and Domain agents, and the real student. Among those entities, there is only one Mediator agent instance and one Domain agent instance formally identified, respectively, by labels M and D. On the other hand, there are as many instances of Learner agent as there are real students using the system. For each student there is a corresponding and specific Learner agent and vice-versa. In this sense, a single identifier L_i is used to reference both the Learner agent and its corresponding real student, with $1 \leq i \leq n$ where n is the number of students using the system. When is clear from the context that there is only one student, only the symbol L will be used to identify this agent.

The knowledge to be shared between agents is modelled by several distinct propositions that represent the beliefs these agents need to communicate one to another. In AMPLIA it was also necessary to model propositions that are probabilistic, that is, propositions that have not only a true or false meaning, but have a subjective probability associated to it. All propositions used in AMPLIA were defined in SLP language. Probabilistic information, as the belief model of a student solution (identified by terms S), is constructed using techniques show in Sect. 6.6.6. The propositions are modelled by formula schemes (or templates). The formal parameters used by these schemes must be considered syntactic meta-variables over specific formulas or terms of SLP. However, because all formulas θ used in $\boldsymbol{BP}(a,\theta)$ terms of SLP must be sentences, when a formula scheme is used as the content of some PACL communicative act then all its formal parameter must be instantiated only to grounded literal terms, transforming the formula scheme in a sentence.

Medical study cases used in the system are represented by SLP terms that are generically identified by the formal parameter CoS. These terms are composed by a informal (textual and graphical) description of the case and an associated list of probabilistic variables necessary to the case.

Following the representation of bayesian networks by SLP formulas, presented in Sect. 6.6.6, the reasoning models that can be stated by the students, as solutions for some study case, are bayesian networks represented by formula schemes (or templates) similar to the formulas presented in Sect. 6.6.6. These formulas will be identified in the following definitions by the formal parameter S. If the instant of time t when the solution was stated is important in the context, it will be indicated by a subscript S_t.

The main classification of these solutions is identified by parameter C. Section 4.4 of Chap. 4 define possible classifications for solutions as: *Not-feasible*, *Incorrect*, *Potential*, *Satisfying* and *Complete*.

Following, Sect. 4.4 of Chap. 4, which defines teaching tactics simply as message texts to be show to students, then these tactics will be represented by *SLP* literal terms identified by the parameter TT. In the same way of identifiers of students' solutions S, if the instant of time t when the tactics should be applied is important, then it will be indicated by a subscript TT_t.

Logical knowledge shared by AMPLIA agents is represented by four different logical propositions: $StudyCase(CoS,L)$, $Sol(CoS,L,S_t)$, $Class(CoS,L,S,C)$ and $TTactic(CoS,L,TT_{t+1})$. The proposition $StudyCase(CoS,L)$ denotes the information about a given case of study that has to be shared between AMPLIA's agents and the student L. It asserts that CoS is the case of study to be presented as the reasoning problem which student L should solve. Proposition $Sol(CoS,L,S_t)$ states that the student L_i believes, at time t, that there is a Bayesian Network model S_t that solves the problem CoS. That proposition is the basic form in which students can inform to the system about the solutions that they are giving for a study case.

The classification and analysis of the student's solutions by the system are expressed through the logical propositions $Class(CoS,L,S,C)$ which tells what is the main classification (C) of the bayesian network model S.

The necessary information to deal with conflicts found in a given solution created by the student are expressed through $TTactic(CoS,L,TT_{t+1})$, which asserts that a new teaching tactics, expressed by TT_{t+1}, has to be applied to help the student to solve the problem CoS in the next cycle of pedagogical negotiation identified by time $t+1$.

The probabilistic knowledge shared by AMPLIA agents is directly related to the probabilistic inferences made by these agents as presented in Sects. 4.3 and 4.4 of Chap. 4. The Mediator agent selects the best tactics for proceeding with the PN process through an Influence Diagram which uses as parameter the results of the network evaluation, made by Domain agent, students' declared confidence, and credibility resulting from the inference the Learner agent made. Both kind of information are expressed as probabilistic propositions: $Conf(CoS,L,S)$ that expresses the degree of confidence that student L has that the model S is the correct solution for the study case CoS and $Cred(CoS,L,S)$ expresses the degree of credibility that Learner agent assigned to the solution S made by student L. Credibility here (from the system point of view) is the degree of belief in a student's autonomous action that can be inferred (by Learner Agent) through the analysis of the student's activity records.

Logical propositions presented in this section express possible beliefs of the agents. These beliefs are shared between agents through **inform** acts. However for some agent a to do the **inform** act about some proposition θ it is required, by the communication theory that fundaments FIPA, that this agent effectively believes in θ, that is, that $\boldsymbol{B}(a,\theta)$ must hold. In the same way, all probabilistic propositions are informed by one agent to another by **inform-bp** acts that also require a similar condition. If some agent a inform another agent that its degree of belief in some proposition θ is p, it is required that $\boldsymbol{BP}(a,\theta)=p$ must be true.

In this way, when some particular communication process, mediated by **inform** acts, is occurring, there always exist a *semantic communication space* (or only a communication space for short) associated to this process. This space is formed by the set of all *SLP* semantic models that satisfy the conditions $\boldsymbol{B}(a,\theta)$ associated to the agents a and sentences θ involved in the communication. This concept also can be generalized to probabilistic propositions communicated by **inform-bp** acts. The communication space in this case is formed by all semantic models that satisfy the set of $\boldsymbol{BP}(a,\theta)=p$ conditions. Because the p values in $\boldsymbol{BP}(a,\theta)=p$ conditions must be probabilities, then set of all semantic models that satisfy these conditions can also be considered as the sample space of the term $\boldsymbol{BP}(a,\theta)$.

6.6.4 Formalization of Initial and Final Conditions of Negotiation

PN processes model all interactions that occur in AMPLIA system, where the main goal is to set and reinforce a high level of confidence among participants of the process. It is not a generic confidence, but a very specific and objective one, related to skills the student has reached and showed with relation to the learning domain. The PN process in AMPLIA should be seen as a way of reducing the initial asymmetry of the confidence relation between teacher and student about the topic studied, maximising the confidence of all. The confidence relationship used in PN is not an absolute (or strong) trustfulness relationship, but it is directly derived from expectations each kind of agent (teacher or student) have with respect to the other agent. It is assumed a weaker notion of trust towards an expectation of future actions of an agent, derived from the confidence notion defined by Fischer and Ghidini [7]. However, it is allowed expectations to have degrees, represented by subjective probabilities. In this way it is possible to define the degree of confidence that the teacher t had that the student s will make some action to solve a particular problem θ (make θ true) through the subjective probability associated to a formula expressing the possibility that agent s do something to solve θ:

$$\boldsymbol{BP}(t,\ (\exists e)(\boldsymbol{Feasible}(e,\ \theta) \wedge \boldsymbol{Agent}(s,\ e)) = p \qquad (6.1)$$

The fact that agent s will find some sequence of actions e that solve the problem θ is expressed by the logical equivalent assertion that exist some sequence e ($(\exists e)$), that really solve the problem θ (make θ feasible - $\boldsymbol{Feasible}(e,\ \theta)$) and is caused by agent s ($\boldsymbol{Agent}(s,\ e)$).

The formal concept of confidence adopted in AMPLIA system is based on (6.1) and defined in (6.2):

$$\boldsymbol{CF}(t,\ e,\ \theta) \equiv_{def} \boldsymbol{BP}(t,\ (\exists e)(\boldsymbol{Feasible}(e,\ \theta) \wedge \boldsymbol{Agent}(s,\ e)) \qquad (6.2)$$

$\boldsymbol{CF}(t,\ s,\ \theta)$ is the confidence degree that the teacher t have that some student s will solve the problem θ. Using this characterisation of confidence it is possible to express the confidence relationships that occur in the AMPLIA's implementation of PN, defining the formal conditions equivalent to the initial and final conditions presented to PN processes in Sect. 6.3. Here it is presented formally only the teacher's side (i.e., the system's side) of these conditions.

To be accepted some solution S must be properly created and declared by the student L and must be complete. These conditions are expressed by:

$$Sol(CoS,L,S) \wedge Class(CoS,L,S,Complete) \quad (6.3)$$

Using the condition (6.3) the initial conditions (IP.1) and (IP.2) defined in Sect. 6.4, can be expressed by the conditions (6.4) and (6.5) bellow, for all study cases *CoS'* similar to the study case *CoS* initially presented to the student:

$$\boldsymbol{CF}(M,\ D,\ Sol(CoS',D,S_D) \wedge Class(CoS',D,S_D,Complete)) \geq \beta \quad (6.4)$$

$$\boldsymbol{CF}(M,\ L,\ Sol(CoS',L,S_0) \wedge Class(CoS',L,S_0,Complete)) \geq \alpha \quad (6.5)$$

The condition (6.4) states that the degree of confidence, which Mediator agent has, that the solution model S_D from Domain agent is complete is greater than or equal to a coefficient β that represent the "high confidence degree" stated in (IP.1). The condition (6.5) states that the degree of belief that Mediator agent M has in the possibility of student L to create an appropriate initial solution S_0 to the study case *CoS'* is greater than or equal to a coefficient α that specify the initial appropriate values for the "low confidence degrees" stated in (IP.2).

The formulation of initial conditions for PN processes, as declared in (6.4) and (6.5), offer an interesting starting point to formalize (IP.1) and (IP.2). They are, however, not entirely appropriate because the similarity notion between study cases *CoS* and *CoS'* used in the definition is not a formal notion. This similarity clearly depends on the "meaning" of these study cases for the students that they are presented. But this meaning is not a formal notion. These cases are only textual and graphical descriptions of reasoning problems that should be understood by the student. The meaning of these descriptions is the intuitive understanding process made by the student. However, the key concept behind this question is that the teacher expects the student be able to solve similar cases to the "test" case *CoS* after have successfully solved *CoS*. From the pedagogical point of view (and from the teacher's point of view) the *CoS* represent a particular set of "external" or "real world" cases that the student is expected to be able to handle, that is, *CoS* is a valid representative for the real world situations related to the topic being studied.

Considering that (IP.1) and (IP.2) are only initial working conditions it is possible to simplify the matter restricting the conditions (6.4) and (6.5) only to the study case *CoS* that represent the topic to be studied[6]:

$$\boldsymbol{CF}(M,\ D,\ Sol(CoS,D,S_D) \wedge Class(CoS,D,S_D,Complete)) \geq \beta \quad (6.6)$$

$$\boldsymbol{CF}(M,\ L,\ Sol(CoS,L,S_0) \wedge Class(CoS,L,S_0,Complete)) \geq \alpha \quad (6.7)$$

When is assumed that exist a previously selected representative case for the topic of study in question, then there is no meaning problem from a

[6] Surelly the teacher should not interrupt the teach-learning process only because the initial expectations are not fullfiled at the beginning of the work with the student.

formal point of view (and from the system point of view). The formal semantics (meaning) associated to some *CoS* depends on *SLP* semantic models. The semantic of a given *CoS* term is given by all elements of the domain's model which can be mapped to this kind of term. It is possible to be sure that appropriate domains and semantic models always exist because *CoS* is a fully grounded literal logical term. In this case, it is possible to prove that appropriate domains and models always exist.

The final conditions (FP.1) and (FP.2) defined in Sect. 6.4, can be analysed in a similar way, using the characterisation of confidence defined in (6.2). The condition (FP.1) is only the condition (IP.1) restated, so the formalization is the same. However, the analysis of condition (FP.2) is a little bit more complex.

In the same way of initial conditions, the final condition (FP.2) can be stated, for all study cases *CoS'* similar to the study case *CoS* and for some time f, as follows:

$$\boldsymbol{CF}(M,\ L,\ Sol(CoS',L,S_f)\ \wedge\ Class(CoS',L,S_f,Complete)) \geq \beta \qquad (6.8)$$

The problem with condition (6.8) is, besides the question of similarity between study cases, is to define clearly how the confidence level of the Mediator agent in the student will be established. Because of the similarity problem, condition (6.8) cannot be used to make a direct formal analysis. What is needed is some other formula or set of formulas that will imply in the condition (6.8).

One obvious necessary requirement to infer condition (6.8) is that the student L can solve the case of study in question in some instant of time, that is, there is a time f and solution S_f where the following condition holds for the study case *CoS* and student L:

$$Sol(CoS,L,S_f)\ \wedge\ Class(CoS,L,S_f,Complete) \qquad (6.9)$$

Note that this condition is not a matter of belief of some agent, but a simple logical condition that must be true or false.

In AMPLIA, it is assumed that the condition (6.9) is necessary but not sufficient to entail logically the condition (6.8). To be able to infer (6.8) it is presupposed that is also necessary to take into account the self-confidence *Conf*(*CoS*,*L*,*S*) expressed by the student and the credibility *Cred*(*CoS*,*L*,*S*) that the system had in the student. In this way, the PN process is successful when *Conf*(*CoS*,*L*,*S*) and *Cred*(*CoS*,*L*,*S*) reach (or surpass) a proper predefined threshold level and when condition (6.9) is satisfied.

The additional successful conditions for PN processes are formalised, for some time f and solution S_f, by the following set of formulas:

$$Conf(CoS,L_i,S_f) \geq \beta_1 \qquad (6.10)$$

$$Cred(CoS,L_i,S_f) \geq \beta_2 \qquad (6.11)$$

The coefficients β_1 and $\beta_2$represent the expected pre-defined threshold levels for, respectively, self-confidence and credibility.

Only when conditions (6.9), (6.10) and (6.11) are satisfied, then AMPLIA consider that the PN process has successfully terminated and that condition (6.8) holds. However, is important to note that formulas (6.6), (6.7), (6.9),

(6.10) and (6.11) are not explicitly declared in the system. Condition (6.6) and the equivalent formula for final condition (FP.1) are basic design assumptions of the implementation of Mediator and Domain agents. Condition (6.7) is not used, that is, the coefficient α is assumed to be 0, so the condition is trivially true. Conditions (6.9), (6.10) and (6.11) are effectively incorporated in the influence diagram (see Chap. 4 Sect. 4.4) that models the decisions made by Mediator Agent. Parameters $\beta_1$and $\beta_2$are implemented as threshold probability parameters of this diagram.

6.6.5 Agent Interaction and Negotiation Space

The formalization of initial and final conditions for PN processes established in previous section. is necessary but is not enough to fully specify in an abstract and formal way the evolution of these kind of processes. What is needed is some model for the intermediary phases of the PN processes. In this section, this model will be built combining the communication spaces related to the interactions that occur between AMPLIA agents with the conditions defined in Sect. 6.6.4.

The resulting model follows the generic framework for automated negotiation presented by Jennings and others [16], in the sense that it is possible to specify the PN process as a searching process in the communication space of the propositions presented in Sect. 6.6.3. In the framework presented in [16], negotiation is described as a distributed search process over potential agreement space (see Fig. 6.2). Dimension and topology of this space are determinate by the structure of negotiation object. For example, the bi-dimensional space presented in Fig. 6.2 should be composed by 2 attributes.

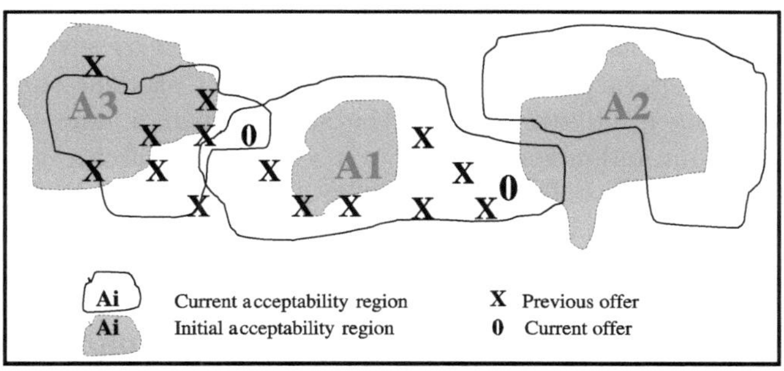

Fig. 6.2. The Negotiation Agreement Space

The initial acceptability regions for agents A1, A2 and A3 are indicated by grey filled areas. In the initial phase of the negotiation process, there are no common agreement spaces, because these regions do not intersect. However, after several set of negotiation offers (indicated by X) these acceptability regions change in a way that a common agreement space appears between agent

A3 and A1. The current acceptability regions are indicated by non-filled areas and currently there is an intersection between acceptability regions of A1 and A3.

In terms of AMPLIA system it is considered that the agreement space is the set of all possible communication spaces that can be constructed for the probabilistic propositions $Conf(CoS,L,S)$ and $Cred(CoS,L,S)$, and for the logical propositions $Sol(CoS,L,S)$ and $Class(CoS,L,S,C)$.

The communication spaces for these propositions define the dimensions and structure of the agreement space. The acceptability regions of the agreement space are defined by formal conditions (6.9), (6.10) and (6.11) defined in Sect. 6.6.4 over these propositions, which represent the final condition (FP.2) for PN processes (see Sect. 6.4). The agreement space of the system is the subset of the acceptability regions that intersect with the acceptability regions of the student.

Note that PN conditions were not defined formally for the student and, consequently, there are no explicitly defined students' acceptability regions of the PN process. However, implicitly it is required that students' acceptability regions intersect with AMPLIA's acceptability regions. The AMPLIA system only will accept (or perceive) solutions in their acceptability regions and will try, through the use of $TTactic(CoS,L,TT_{t+1})$ propositions, to reach to some of these regions.

In being so, for some particular case of study CoS and student L both the agents in AMPLIA and the real student will try to find out some particular appropriate solution model S_f in a time f, that is, they need to find a appropriate point in the agreement space. To be accepted, the solution S must be properly created and declared by the student through $Sol(CoS,L,S_f)$ and the probabilities assigned to $Conf(CoS,L,S_f)$ and $Cred(CoS,L,S_f)$ must reach an appropriate value in the agreement space.

The objective of the negotiation process is achieved by several teach and learning cyclical interactions that occur between the student and the agents of AMPLIA. One cycle starts when the student had a new solution to a particular problem and generally follows a sequence of interactions involving first Learner Agent, then Domain Agent and the Mediator Agent, returning to the Learner Agent and the student.

One particular cycle t of interaction between Learner agent and Domain agent starts when a new solution $Sol(CoS,L,S_t)$ is built by the student and sent to the Domain agent through a communicative act **inform**, which is autonomously emitted by the Learner agent. Besides this, the Learner agent also uses **inform-bp** acts to inform $Conf(CoS,L,S_t)$ and $Cred(CoS,L,S_t)$. Domain agent analyses the student's solutions and send its conclusions to Mediator agent through $Class(CoS,L,S_t,C)$ logical propositions. Finally, Mediator agent based on this information defines if a new teaching tactics should be presented to the student. The tactic, represented by $TTactic(CoS,L,TT_{t+1})$, it is sent directly to the Learner agent through an **inform** act when the Mediator agent decides that a new teaching tactics must be adopted in the next cycle *t+1*.

The PN cycles for a particular case of study and student will have a successful conclusion when *Conf*(CoS,L,S_f), *Cred*(CoS,L,S_f), *Sol*(CoS,L,S_f) and *Class*(CoS,L,S_f,C) reach appropriate values for some final cycle f. However, students can stop these cycles at any time if they decide to start to study a new case.

6.6.6 Representation of Probabilistic Knowledge

With *SLP* is possible to represent the bayesian networks and inference diagrams used by AMPLIA's agents, including all inference processes associated to these networks. Bachus [1] has shown how simple bayesian networks can be represented in probabilistic logics. Gluz [11] extended this representation scheme showing how any discrete bayesian network can be transformed in the equivalent set of *SLP* formulas and how partitioned bayesian networks (MSBNs) and associated consistency maintenance protocols can also be represented by SLP and PACL. However, it is important to note that these logical representation formats for bayesian networks do not have the computational efficiency that pure BN inference methods have. Nevertheless, the main point in the logic representation of BN is its declarative character that is precisely defined by a formal axiomatic semantic. Because the intended use for this representation is in the communication of some BN from one agent to another, then it is believed that this precise, declarative and axiomatic character are important assets when defining the structure and meaning of some knowledge that has to be shared among two or more agents. This is true for agents that exchange logical knowledge and should be true when the exchange of probabilistic knowledge is in question.

First question to be considered when trying to represent BNs by *SLP* formulas is how probabilistic variables (nodes) should be represented. Probabilistic variables can range over distinct values (events) in a *sample space*. One basic assumption it is that events from the sample space can be represented by elements of the SLP domain. In this way, it is possible to use unary logical predicates to represent these variables.

Bayesian networks adopt the subjective interpretation of the probability concept (they are also known as *beliefs networks*), so they can be appropriately represented by the $\boldsymbol{BP}(a,\varphi)$ probabilistic terms. Arcs between the nodes (variables) of the network are interpreted as conditional probabilities between the variables corresponding to the nodes. The conditional probability operator $\boldsymbol{BP}(a, \varphi \mid \psi)$ defined in *SLP* is used to represent the arcs.

Finally, an axiom schema of SLP can represent the equation that defines the properties of the Joint Probability Distribution (JPD) function of the bayesian network.

A Bayesian Network (BN) is essentially a graphical device to express and calculate the JPD function of some domain that is not affected by the

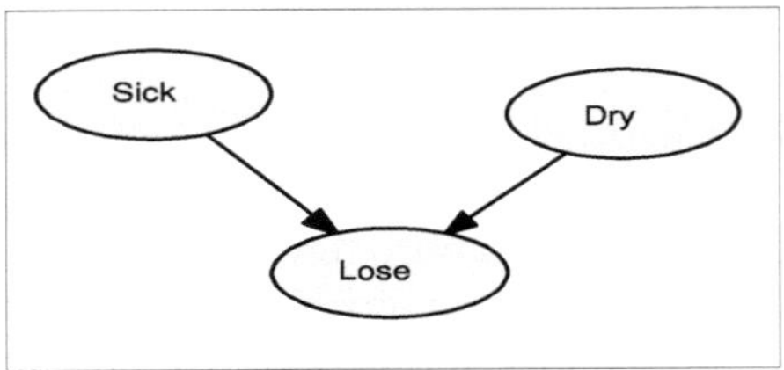

Fig. 6.3. Example of Bayesian Network

combinatorial explosion problem. In this section it is used the example BN presented in Fig. 6.3, extracted from HUGIN system tutorial [15]. The BN expresses how can be estimated the conditional probability of some tree to be loosening leaves, if it is sick and the weather is dry. Figure 6.3 shows qualitative relationships between probabilistic variables *Sick*, *Dry* and *Lose*. Table 6.2 shows the quantitative model associated to this BN that expresses previous probabilities for some tree to be sick ($p(Sick)$) or for dry weather ($p(Dry)$) occurs. It also defines the conditional probability that a tree looses its leaves if it is sick and weather is dry ($p(Lose|Sick,\ Dry)$).

Table 6.2. Conditional Probabilities for Example BN

				$p(Lose \mid Sick,\ Dry)$				
				Sick	sick		not	
Sick	$p(Sick)$	*Dry*	$p(Dry)$	*Lose**Dry*	dry	not	dry	not
sick	0.1	dry	0.1	Yes	0.95	0.9	0.85	0.02
not	0.9	not	0.9	Not	0.05	0.1	0.15	0.98

Variables from example BN will be represented by predicates **Lose**, **Sick**and **Dry**:

$$(\forall x)\ (\mathbf{Sick}(x) \rightarrow (x{=}\text{sick}) \vee (x{=}\text{not})) \wedge \qquad (6.12)$$
$$(\forall x)\ (\mathbf{Dry}(x) \rightarrow (x{=}\text{dry}) \vee (x{=}\text{not})) \wedge$$
$$(\forall x)\ (\mathbf{Lose}(x) \rightarrow (x{=}\text{yes}) \vee (x{=}\text{not}))$$

The quantitative component of the BN, which informs the strengths of the causal influences from its parent nodes, can be fully specified by the following equations, for any agent a:

$$\boldsymbol{BP}(a,\ \mathbf{Sick}(\text{sick}))=\mathbf{0.1} \wedge \boldsymbol{BP}(a,\ \mathbf{Sick}(\text{not}))=\mathbf{0.9} \wedge \qquad (6.13)$$
$$\boldsymbol{BP}(a,\ \mathbf{Dry}(\text{dry}))=\mathbf{0.1} \wedge \boldsymbol{BP}(a,\ \mathbf{Dry}(\text{not}))=\mathbf{0.9} \wedge$$
$$\boldsymbol{BP}(a,\ \mathbf{Lose}(\text{yes}) \mid \mathbf{Sick}(\text{sick}) \wedge \mathbf{Dry}(\text{dry}))=\mathbf{0.95} \wedge$$
$$\boldsymbol{BP}(a,\ \mathbf{Lose}(\text{not}) \mid \mathbf{Sick}(\text{sick}) \wedge \mathbf{Dry}(\text{dry}))=\mathbf{0.05} \wedge$$
$$\boldsymbol{BP}(a,\ \mathbf{Lose}(\text{yes}) \mid \mathbf{Sick}(\text{sick}) \wedge \mathbf{Dry}(\text{not}))=\mathbf{0.9} \wedge$$
$$\boldsymbol{BP}(a,\ \mathbf{Lose}(\text{not}) \mid \mathbf{Sick}(\text{sick}) \wedge \mathbf{Dry}(\text{not}))=\mathbf{0.1} \wedge$$
$$\boldsymbol{BP}(a,\ \mathbf{Lose}(\text{yes}) \mid \mathbf{Sick}(\text{not}) \wedge \mathbf{Dry}(\text{dry}))=\mathbf{0.85} \wedge$$
$$\boldsymbol{BP}(a,\ \mathbf{Lose}(\text{not}) \mid \mathbf{Sick}(\text{not}) \wedge \mathbf{Dry}(\text{dry}))=\mathbf{0.15} \wedge$$
$$\boldsymbol{BP}(a,\ \mathbf{Lose}(\text{yes}) \mid \mathbf{Sick}(\text{not}) \wedge \mathbf{Dry}(\text{not}))=\mathbf{0.02} \wedge$$

$\boldsymbol{BP}(a, \textbf{Lose}(\text{not}) \mid \textbf{Sick}(\text{not}) \wedge \textbf{Dry}(\text{not})) = \mathbf{0.98}$

Finally, the JPD of this BN can be fully specified by the axiom schema bellow:

$$\boldsymbol{BP}(a, \textbf{Lose}(t_1) \wedge \textbf{Sick}(t_2) \wedge \textbf{Dry}(t_3)) = \boldsymbol{BP}(a, \textbf{Lose}(t_1) \mid \textbf{Sick}(t_2) \wedge \textbf{Dry}(t_3)) \times \boldsymbol{BP}(a, \textbf{Sick}(t_2)) \times \boldsymbol{BP}(a, \textbf{Dry}(t_3)) \tag{6.14}$$

Axiom schema (6.14) defines $\boldsymbol{BP}(a, \textbf{Lose}(t_1) \wedge \textbf{Sick}(t_2) \wedge \textbf{Dry}(t_3))$ as a function that can be used to calculate the probabilities of all combination of values from probabilistic variables. Symbols t_1, t_2, and t_3 represent meta-variables that can be substituted by terms of *SLP*. In *SLP* is not allowed free variables in formulas under the scope of a ***BP*** term [11, 12], then t_1, t_2, and t_3 meta-variables from axiom schema (6.14), can be substituted only by completely grounded atoms. In the example, because of condition (6.12), only valid terms are sick, dry, yes and no. The formula above also encodes the structural component of the BN, because the conditional probability operator $\mid$ in $\boldsymbol{BP}(a, \textbf{Lose}(t_1) \mid \textbf{Sick}(t_2) \wedge \textbf{Dry}(t_3))$ can be interpreted as stating that **Sick** and **Dry** nodes are connected to the **Lose** node.

The axiom scheme (6.14) and formulas (6.12) and (6.13) represent completely the example BN because the probability of any combination of instantiations of t_1, t_2 and t_3 allowed by condition (6.12) can be calculated (inferred) through the application of the probabilistic axioms and inference rules of *SLP*. These formulas are not a simple "encoding" of the example BN in a logical language, because it is possible to "extract" (to infer) using logical deduction all knowledge usually implied by this BN. For example the expression $p = \mathbf{BP}(a, \textbf{Lose}(\text{yes}) \wedge \textbf{Sick}(\text{sick}) \wedge \textbf{Dry}(\text{not}))$ denotes the subjective probability p assigned by some agent a that its tree loses their leaves, if it is sick but the weather it is no dry. This probability p can be discovered simply substituting appropriate values on axiom schema (6.14):

$$p = \boldsymbol{BP}(a, \textbf{Lose}(\text{yes}) \mid \textbf{Sick}(\text{sick}) \wedge \textbf{Dry}(\text{not})) \times \boldsymbol{BP}(a, \textbf{Sick}(\text{sick})) \times \boldsymbol{BP}(a, \textbf{Dry}(\text{not}))$$

However, by equations defined in (6.13), the value of p can be calculated as follows:

$$\begin{aligned} p &= \boldsymbol{BP}(a, \textbf{Lose}(\text{yes}) \mid \textbf{Sick}(\text{sick}) \wedge \textbf{Dry}(\text{not})) \times \boldsymbol{BP}(a, \textbf{Sick}(\text{sick})) \times \boldsymbol{BP}(a, \textbf{Dry}(\text{not})) \\ &= \boldsymbol{BP}(a, \textbf{Lose}(\text{yes}) \mid \textbf{Sick}(\text{sick}) \wedge \textbf{Dry}(\text{not})) \times \mathbf{0.1} \times \mathbf{0.9} \\ &= \mathbf{0.9} \times \mathbf{0.1} \times \mathbf{0.9} = \mathbf{0.081} \end{aligned}$$

This value is identical to the obtained in HUGIN system for the same case.

6.7 Pedagogical Negotiation and the Real World

The medical student must have the opportunity to build diagnostic models of diseases, including probable causes, associated symptoms and must be able to assess the application of a model. Thus, the student has the opportunity to apply action strategies while creating diagnostic reasoning. Medical teaching usually uses resources such as cases, topics or articles discussion during classroom seminars. Computer science resources, such as discussion lists, teleconference and chats are used for communication in distance education. The number of learning environments that use computer science resources increases each day, such as decision support systems and Intelligent Tutoring Systems. They try to support the learning process according to different pedagogic lines.

The physician, for example, can diagnose a disease based on some symptoms, but this diagnostic is only a hypothesis, because it can be wrong. Such an error can be linked to the incomplete knowledge of the pathology approached, if determinant symptoms were not detected due to the progression of the disease, which is in its onset phase. Even thus, this diagnosis is more reliable than a simple guess. Currently, to handle uncertainty the medical community has been provided with decision support systems based on probabilistic reasoning. These systems can be consulted by professionals and, sometimes, used as pedagogic resources. This is the point where academic difficulties are found: a student can consult one of these systems and reproduce the expert's diagnostic hypothesis, but this does not warrant that the student will come to understand all its complexity and, much less, that the student will be able to do the diagnostic based on another set of variables. The ideal would be to use all the expert's hypotheses (provided it was possible) so that the student would be able to understand how and why a given diagnostic was selected. In other words, it is important that the student be conscious about the entire process that is involved in the construction and selection of a hypothesis and not only about the outcome of this process. The main goal for such student is, on top of making a correct diagnosis, to understand how different variables (clinical history, symptoms, and laboratory findings) are probabilistically related among each other.

The challenge was to create a learning environment that could really use the key concepts embedded in the idea of negotiation in a teaching-learning process (pedagogic negotiation), aiming at setting the project's principles, which are: *Symmetry between man and machine* and *existence of negotiation spaces*. The relationship between the user and a system is usually not symmetric. For example, decision support systems can make decisions regardless of users, only considering its knowledge and the inference to request data or generate explanations. The challenge is in the search of symmetry between man and machine. Such symmetry provides the same possibilities to the user and the system as to actions, and symmetric rights for decision taking. For a give interaction of tasks, the negotiation behaviour among agents and their power will be in a great part determined by the differences on knowledge about the domain.

In an asymmetry mode, an agent has always the decisive word; there is no space for a real negotiation. In the symmetric mode, there is no pre-defined winner; conflicts need to be solved through a negotiation. Precisely, the cognitive processes that trigger an explicit declaration, justify an argument or refuse the partner's point of view are most likely to explain the reasons why collaborative learning is more efficient than lonely learning.

The main functions of the interactive learning environments (explanation, education, diagnostic) are traditionally implemented as one-way mechanisms; this means the system has the total control. AMPLIA, however, tries to treat them as bilateral processes, this means that a diagnostic model (Bayesian network) is not built by the student and sent to the system. This network is built collaboratively, and there are some negotiation moments. The quality of such diagnosis depends specially on the collaboration between the student and the system. It is well known that even in education, the student performs an active role that helps teachers to tune their interventions.

It is clear that for a negotiation to take place there must be a level of latitude available to agents, otherwise anything can be negotiated. This defines the global negotiation space within which the two agents try to build a shared understanding of the problem and its solution. The negotiation naturally does not take place in a single plan. Dillenbourg et al. [5] say that human partners do not negotiate a single shared representation, but they really develop several shared representations, that is, they move in a mosaic of different negotiation spaces. These spaces differ according to the nature of the information that will have to be negotiated and for the negotiation mechanisms. In the human-machine collaboration, more precisely in collaborative learning environments, usually a single negotiation space takes place.

Due to the domain knowledge chosen in AMPLIA – medical, which requires uncertain reasoning – the project has a series of features that allow for the consolidation of an effective pedagogic negotiation process. AMPLIA is a collaborative system that provides a strong symmetry between human and artificial agents in terms of action.

Then, the negotiation object in AMPLIA is the belief on a diagnostic hypothesis for a given clinical case that the student expresses by using an editor of Bayesian networks. As to the pedagogic aspect, the Bayesian networks are ideal tools for the construction of uncertain knowledge, which makes the negotiation space wider and more symmetric.

The use of Bayesian networks as a tool to interact with the student rules out the constraints of the system competence to understand language. The process of a Bayesian model construction is always followed by the system, which allows for an intervention whenever it is necessary, that is, at each interaction cycle the system accepts or rejects the learner's Bayesian model, using adequate arguments for every situation.

Figure 6.4 shows (first window screen) the interface of AMPLIA collaborative editor the student used for the construction of the diagnostic hypotheses. The second window presents the Bayesian network built by the expert.

The argumentative resources of the system (pedagogic strategies and tactics) try to make students reflect upon their actions (construction and

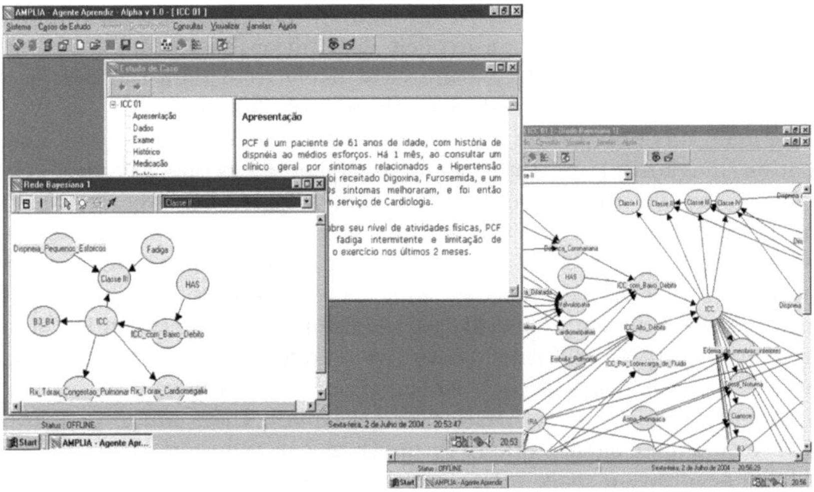

Fig. 6.4. AMPLIA collaborative editor interface

alteration of a Bayesian model), aiming at improving their diagnostic hypothesis. The form of argumentation offered to students is related to the liberty to modify their Bayesian model or not, and the declaration of their self-confidence. It is important to highlight that, although learner and system's arguments are different, the interaction language is the Bayesian Network, easily understood by the student and the system. This means that the system has total conditions of understanding and assessing the network, supported on the Bayesian Network built by the expert and in a database with real cases. Thus, the system intuitively offers to students the opportunity to criticize the expert's model, when the students' performance (right identification of the diagnostic of real cases) is higher than expert's network performance.

AMPLIA was built following an evolutionary strategy with two big development phases. In the first phase, were developed prototypes for the three types of agents, specified the initial model for their communication and finally implemented the communication tasks in each kind of agent. The implementation was based in heterogeneous and distributed agent architecture [9]. Communication between agents was implemented by FIPA-ACL acts (see Sect. 6.6.1), extended to support probabilities as described in Sect. 6.6.2. All contents, including BN models and logical propositions, exchanged in these acts were represented in a format derived from the BN representation format presented in Sect. 6.6.6. The formalization of PN process presented in Sects. 6.6.3, 6.6.4 and 6.6.5 provided the design guidelines for the implementation of the communication tasks of all AMPLIA's agents.

The following phase took place in a real teaching environment, through the experimentation and refinement of AMPLIA in an extension course for the medical area realized at the Hospital de Clínicas de Porto Alegre. The refinement and adjustment of the system was performed in a gradual way, so that it could be available since the start of the project, although not in its full capacities. The course comprised two modules: the first approached pedagogic resources, theoretical concepts on uncertain domains, probabilistic networks and knowledge representation. In the second module, the teachers built the expert's networks that were incorporated into the Domain Agent knowledge base. Eventually, groups of students used AMPLIA for the first assessments and validations.

The sixteen health professionals enrolled at the course came from different medical areas, such as surgery, anaesthesia, psychiatry, endocrinology, cardiology, internal medicine, odontology and nursery. The course aimed at making the environment known for the medical teaching community, once the design of the environment was aimed at such professionals. The goal was not only to validate the teaching-learning environment and its formal model, but also to make physicians aware that such tool can be used as an additional tool in the qualification of young physicians.

Such an interaction and collaboration methodology had a fundamental role in the development process of the project. It allowed for the constant evaluation of the environment features in several aspects (formal education, culture, competences, preferences), but keep it aligned to basic ILE objectives: (a) learning – in this case the student-teacher interaction mediated by the machine, and (b) continued education – up to date information is accessed through the machine.

6.8 Summary

This chapter discussed the pedagogical negotiation process (PN) involved in the implementation of a real learning environment – AMPLIA. It focuses on the formal aspects of the negotiation process, trying to extract the most general characteristics of this process. The application of formal approaches to understand or conceptualise aspects of educational processes is a relatively new research area, at least in the sense of the formalisation of Intelligent Tutoring Systems.

The negotiation is characterized by: i) the negotiation object (belief on a knowledge domain), ii) the negotiation initial state (absence of an agreement, which is characterized by an unbalance between credibility, confidence, and a low BN model quality); iii) the final state (highest level of balance between credibility and confidence, and good BN model quality); and iv) the negotiation processes (from state ii to state iii). This is the base of the negotiation model developed in AMPLIA.

The goal of the negotiation model is to increase, as much as possible: (a) the performance of the model the students build; (b) the confidence that teachers and tutors have in the students' ability to diagnose cases; and the students' confidence on their own ability to diagnose cases; and (c) the students' confidence on their own ability to diagnose diseases

During AMPLIA development it has been studied if the use of BNs as a pedagogical resource would be feasible, if they would enable students to model their knowledge, follow students actions during the learning process, make inferences through a probabilistic agent, and select pedagogical actions that have maximum utility for each student at each moment of the knowledge construction process. All these applications are assumed probabilistic, as they involve all the complexity and dynamics of a human agent learning process, but with the possibility of being followed by artificial agents.

The challenge was to create a learning environment that could really use the key concepts embedded in the idea of negotiation in a teaching-learning process (pedagogic negotiation), aiming at setting the project's principles, which are: Symmetry between man and machine and existence of negotiation spaces.

The implementation of AMPLIA was performed in a gradual way, so that it could be available since the start of the project, although not in its full capacities. The system was tested at the Hospital de Clínicas de Porto Alegre in an extension course that took place in the hospital. The course comprised two modules: the first approached pedagogic resources, theoretical concepts on uncertain domains, probabilistic networks and knowledge representation. In the second module, the teachers built the expert's networks that were incorporated into the Domain Agent knowledge base.

The results obtained in these preliminary tests have shown a convergence with the observations carried out by the teacher who followed the students during the process of network construction. This means that the teacher probably would use tactics and strategies similar to those selected by the system, to mediate the process. Summing up, the student model the teacher elaborated is similar to the model constructed in the AMPLIA environment and the decision taken by the environment is compliant with the teacher pedagogical position.

As future works it is intended to make AMPLIA available over the Web and to refine the graphic editor so that it can allow for a simultaneous work of several students in the same case of study. The student's self-confidence declaration can also be approached as a future work, focusing the student's emotions, which were not considered in the present phase.

Acknowledgments. The authors gratefully acknowledge the Brazilian agencies CAPES, CNPq and FAPERGS for the partial support to this research project.

References

1. Bacchus, F. (1990) Lp, a Logic for Representing and Reasoning wih Statistical

Knowledge. Computational Intelligence. v. 6, p. 209-301, 1990. Available at: <http://citeseer.nj.nec.com/>. Accessed on: Nov. 2003.
2. Baker, M.J. (1994). A model for negotiation in teaching-learning dialogues Journal of Artificial Intelligence in Education, 5 (2), 199-254.
3. Cohen, P.; Levesque, H. (1990) Intention Is Choice with Commitment. Artificial Intelligence, n. 42, p. 213-261, 1990.
4. Dillenbourg, P., Self, J.A. (1992) A framework for learner modelling, Interactive Learning Environments 2, 1992, p.111–137.
5. Dillenbourg, P. et al. (1995) The evolution of research on collaborative learning. In: Spada, E.; Reitman, P. (Ed.). Learning in Humans and Machine: Towards an interdisciplinary learning science. Oxford: Elsevier. 1995. p. 189-211.
6. FIPA. (2002) FIPA Communicative Act Library Specification. Standard SC00037J, FIPA, Dec. 2002. Available at: <http://www.fipa.org/>. Accessed on: Feb. 2004.
7. Fischer, M., Ghidini, C. (2002). The ABC of Rational Agent Modelling. In: Proceedings of AAMAS 2002, Bologna, Italy, 849-856, 2002.
8. Flores, C.D.; Seixas, L.; Gluz, J.C.; Vicari, R.M. A Model of Pedagogical Negotiation. (2005) In: Multi-Agent Systems: Theory and Applications Workshop - MASTA, Covilhã. 12th Encontro Português de Inteligência Artificial (EPIA 2005, LNAI 3808). Berlin:Springer Verlag, v. 1, p. 488-499.
9. Flores, C.D., Seixas, L., Gluz, J.C., Patrício, D., Giacomel, F., Gonçalves, L., Vicari, R.M. (2005) AMPLIA Learning Environment Architecture. In: 13th International Conference on Computers in Education (ICCE2005), Cingapura. Towards Sustainable and Scalable Educational Innovations Informed by the Learning Sciences. Tokyo:IOS Press, p. 662-665.
10. Flores, C.D., Gluz, J.C., Seixas, L., Viccari, R.M. (2004) Amplia Learning Environment: A Proposal for Pedagogical Negotiation. In: Proceedings of 6th International Conference on Enterprise Information Systems. Porto, Portugal, INSTICC, Vol. IV, p:279-286.
11. Gluz, J. C. (2005) Formalization of the Communication of Probabilistic Knowledge in Multiagent Systems: an approach based on Probabilistic Logic (In Portuguese). PhD Thesis. Instituto de Informática, UFRGS, Porto Alegre, 2005
12. Gluz, J. C., Vicari, R. M., Flores, C. D. and Seixas, L. (2006) Formal Analysis of the Communication of Probabilistic Knowledge. Proceedings of IFIP - WCC - AI 2006. Santiago, Chile, 2006.
13. Halpern, J. Y. (1990) An Analysis of First-Order Logics of Probability. Artificial Intelligence, 46: 311-350, 1990.
14. Heckerman, D.; Mamdani, A.; Wellman, M. (1995) Real-world Applications of Bayesian Networks: Introduction. Communications of the ACM, Vol. 38, n. 3. ACM, New York.
15. Hugin. (2005) Available at: <http://www.hugin.com> Accessed on: Aug. 2005.
16. Jennings, N.R., Faratin, P., Lomuscio, A.R., Parsons, S., Sierra, C., Wooldridge, M. (2000) Automated Negotiation: Prospects, Methods and Challenges. In: Int Journal of Group Decision and Negotiation, 2000.
17. Rao, A.S.; Georgeff, M.P. (1991). Modeling rational agents within a BDI-architecture. In: Fikes, R.; Sandewall, E. (eds.) Proceedings of Knowledge Representation and Reasoning (KR&R-91), p. 473-484. San Mateo, CA: Morgan Kaufmann Publishers, 1991.
18. Russel, S., Norvig, P. (1995) Artificial Intelligence: a modern approach. New Jersey: Prentice Hall, 1995.

19. Sadek, M. D. (1992) A Study in the Logic of Intention. In: Procs. of KR92, p. 462-473, Cambridge, USA, 1992.
20. Sandholm, T. W. (1999) Distributed Rational Decision Making. In: Weiss, G. (ed.) Multiagent Systems: A Modern Approach to Distributed Artificial Intelligence. Cambridge: The MIT Press, p.79-120, 1999.
21. Schwarz, B.B.; Neuman, Y.; Gil, J.; Ilya, M. (2001). Effects of argumentative activities on collective and individual arguments. European Conference on Computer-Supported Collaborative Learning – Euro-CSCL 2001, Maastricht, 22 - 24 March 2001.
22. Self, J.A. (1990) Theoretical Foundations for Intelligent Tutoring Systems. In: Journal of Artificial Intelligence in Education, 1(4), p.3-14.
23. Self, J.A. (1992) Computational Viewpoints. In Moyse & Elsom-Cook, pp. 21-40
24. Self, J.A. (1994) Formal approaches to student modelling, in: G.I. McCalla, J. Greer (Eds.), Student Modelling: The Key to Individualized Knowledge-Based Instruction, Springer, Berlin, 1994, p. 295–352.
25. Vicari, R.M., Flores, C.D., Seixas, L., Silvestre, A., Ladeira, M., Coelho, H. (2003) A Multi-Agent Intelligent Environment for Medical Knowledge. In: Journal of Artificial Intelligence in Medicine, Vol.27. Elsevier Science, Amsterdam, p. 335-366.
26. Varian, H. R. (2003) Intermediate Microeconomics: a Modern Approach. W.W. Norton & Company, 2003.
27. Yokoo, M., Ishida, T. (1999) Search Algorithms for Agents. In: Weiss, G. (ed.) Multiagent Systems: A Modern Approach to Distributed Artificial Intelligence. Cambridge: The MIT Press, p.165-200, 1999.

7

A New Approach to Meta-Evaluation Using Fuzzy Logic

Ana C. Letichevsky[21], Marley Vellasco[2], Ricardo Tanscheit[2], and Reinaldo C. Souza[2]

[1] Cesgranrio Foundation, Brazil
anacarolina@cesgranrio.org.br
[2] Department of Electrical Engineering, PUC-RJ, Brazil, (marley, ricardo, reinaldo)@ele.puc-rio.br

Assuring the quality of evaluation is a great challenge to evaluators. The evaluation of an evaluative process is called meta-evaluation. In Brazil there is a great concern about the evaluation quality, although the concept of meta-evaluation is a new one. Standards that a true evaluation should attend, in view of those defined by the Joint Committee on Standards for Educational Evaluation, are currently under discussion. This study presents a new methodology developed in Brazil for meta-evaluation that makes use of fuzzy sets and fuzzy logic concepts. The methodology is composed of an instrument for data collection (checklist for the meta-evaluation of Programs/Projects) and of a fuzzy inference system for treatment of data related to the meta-evaluation of projects and programs. The implementation of a quality meta-evaluation process requires an appropriate methodology for the data collection and the inference process. The main advantages of developing a methodology based on fuzzy logic are: (i) the possibility of working with linguistic rules; (ii) the utilization of an adequate tool to work with the intrinsic imprecision involved in the evaluation of the standards; (iii) facility to incorporate the subjective knowledge of specialists; (iv) facility of adapting the inference process to specific situations. This new methodology makes use of a hierarchical Mamdani-type inference system that comprises 36 rule bases organized in three levels: *standard* (level 1), *category* (level 2), and *meta-evaluation* (level 3). The main advantages of the proposed system are related to: (i) the data collection instrument, which allows intermediary answer; (ii) the inference process and its capacity to adapt to specific needs; and (iii) transparency, by the use of linguistic rules that facilitates the understanding and discussion of the whole process. It is believed that this methodology will facilitate the meta-evaluation process especially in developing countries, where usually there is a significant

Ana C. Letichevsky et al.: *A New Approach to Meta-Evaluation Using Fuzzy Logic*, Studies in Computational Intelligence (SCI) **44**, 147–165 (2007)
www.springerlink.com

lack of qualified professionals in the area and a huge demand for processes of meta-evaluation. This study intends to be a contribution to evaluation as a subject and to meta-evaluation practice.

7.1 Introduction

Currently, evaluation is understood as a process through which a value judgment is made about the focus or object that is being evaluated. Human beings are capable of evaluating people, groups, institutions, organizations, processes and a series of other items. Whenever a value judgment is made, it is necessary to make use of one or more criteria, even if they are not explicit or clear, as often occurs in informal evaluations. This has been done for centuries, before the creation of formal procedures for conducting an evaluation, and is still done nowadays, since we were born evaluators, though not necessarily good evaluators [1].

Since the beginning of the twentieth century, evaluation has been carried out by means of techniques, procedures and systematic methods that make possible to conduct a formal evaluation process. Throughout a century of studies, evaluation has evolved and begun to be understood as a documented process of systematic collection of data and precise, relevant information capable of answering previously asked questions specifying the criteria of excellence. These criteria enable value judgments regarding (i) the merit of the focus of attention (in terms of the internal quality of its components and their functioning) and (ii) the relevance - which refers to its effects, in terms of its impact [2]. Professional evaluation is defined as the systematic determination of the quality of the value of something [3].

Assuring the quality of evaluation is a great challenge to evaluators. The evaluation of an evaluative process is called meta-evaluation. Meta-evaluation is the mechanism used nowadays to face this challenge. The discussions about meta-evaluation carry as a main focus the excellence criteria for an evaluation. This is the starting point for obtaining a quality meta-evaluation, but it does not exhaust the subject. It is necessary to go beyond, to overcome the present limits and find new methodologies that permit the conduction of meta-evaluation processes in more flexible ways, supplying a precise and timely answer, in order to break the geographic limits and permit the conduction of meta-evaluation processes in different countries.

Meta-evaluation can be carried out in different ways, but frequently checklists [4] are used as an instrument for collecting data. Checklists are instruments with items or assertions about a specific focus with options of closed answers. In case of a meta-evaluation process, those assertions must investigate the presence of each one of the standards of a true evaluation in the target evaluation process. Traditionally, data are collected and treated based on classic logic, where the frontier between a point of the scale and another one is always precise. However, it is not easy for a human being to define this frontier precisely, since it may be of a fuzzy nature. In Fuzzy Set Theory [5], a given element can belong to more than one set with different grades of membership (values in the interval [0,1]). The same difficulty that exists in the step of data collection is also faced in the treatment of information, since, in order to perform an evaluation, excellence criteria must be established. These criteria serve to develop a value judgment and can form a support for rules, generally supplied by specialists that are used to verify whether or not the result meets a certain criterion.

When traditional logic is employed in meta-evaluation, the accuracy of the meta-evaluation can be questioned: are we actually measuring what we want to? In fact, experimental results indicate that the context and fuzzy reasoning are typical of half the population [6].

This study presents a methodology proposed and developed in Brazil for meta-evaluation that makes use of the concepts of fuzzy sets and fuzzy logic. This allows for the use of intermediate answers in the process of data collection. In other words, instead of dealing with crisp answers ("accomplished" or "not accomplished", for example), it is possible to indicate that an excellence criterion was partially accomplished in different levels. The answers of this instrument are treated through the use of a Mamdani-type inference system [7], so that the result of the meta-evaluation is eventually obtained. Therefore, the proposed methodology allows: (i) the respondent to provide correct answers that indicate his (her) real understanding with regard to the response to a certain standard; (ii) to use linguistic rules provided by specialists, even with contradictory thinking; (iii) to deal with the intrinsic imprecision that exists in complex problems such as the meta-evaluation process.

7.2 Meta-evaluation

7.2.1 The Concept

When an evaluative process is designed, several aspects, such as evaluative questions, methods and techniques of data collection and identification of the respondents, should, from the beginning, be negotiated with whomever is in charge of the evaluation and with the representatives of those being evaluated. Evaluators and clients must be aware of the bias in the evaluative process, and seek to minimize it whenever possible; when this is not possible, they must report it. After all, meta-evaluations are carried out so that the bias is minimized and the quality of the evaluative process in all its stages is ensured. This includes decisions concerning the execution of the evaluation, the definition of its purpose, design, information collection and analysis, elaboration of budget and contract, management, setting up of the team.

In the same fashion that it is recommended that evaluations be carried out in the formative and summative perspectives [8], meta-evaluations should also be carried out having in view those two perspectives, which up to a certain point complement each other. The formative meta-evaluation is conducted along the evaluative process to improve on the evaluation. Ideally, its starting point should coincide with that of the evaluation. The main objective is to provide the team responsible for carrying out the evaluative process with useful information, in order to improve the process while it is still in progress. The summative meta-evaluation is carried out at the end of the evaluation, in search of conclusive answers about its merit and relevance to those who ordered it, as well as to users and others interested in the process. The aim is to give credibility to the evaluation and to the final results generated by it. In other words, while the role of the formative meta-evaluation is to improve the evaluative process throughout its development, the role of the summative evaluation is to give an account to those involved and to the community at large, and to contribute to the improvement of future processes. Similarly to what occurs in evaluative processes, meta-evaluations may be carried out through the use of different evaluative approaches such as utilization-focused [9, 10], responsive [11, 12], connoisseurship [13, 14] above others.

The concern with the creation of standards (principles obtained through consensus among the people involved in the practice of evaluation, which, if achieved, will guarantee the quality of an evaluation) that the evaluative process should follow is an old one and, perhaps, as old as the concern with evaluation itself. This is a hard task, not only on account of the technical difficulty inherent to it, but also, above all, as a result of the difficulty to sensitize, mobilize, and reach consensus among different pertinent people, so as to produce a technically good work, accepted by those who carry out the evaluation, those who are evaluated, and those who make use of it.

In the sixties, some renowned evaluators, such as Scriven, Stake, and Stufflebeam, began discussing formal procedures and criteria for meta-evaluation, suggesting how good and poor evaluations would be like [15]. However, with the emergence of different lists of criteria for the assessment of the evaluative processes quality, those who order evaluations, those who are evaluated, and those who evaluate, began to question themselves as to which list would be used. In 1975, the Joint Committee on Standards for Educational Evaluation was formed and generated the Standards for Evaluations of Educational Programs, Projects and Materials [16]. Later on, this committee revised the standards developed specifically for the school context and adapted them to personnel evaluation [17] and then to respond to evaluations of any nature. This work resulted in a group of thirty universal standards that should, whenever applicable, be present in the execution of evaluative processes in any area of knowledge [18] and are organized in four characteristic attributes (or categories): utility, feasibility, propriety, and accuracy. Before the beginning of any evaluative process, evaluators and other persons involved in the evaluative process must discuss which standards will guide the evaluation. It is important to point out that there are other less known standards besides those developed by the Joint Committee on standards for Educational Evaluation. Nevertheless, in this paper only those are considered.

7.2.2 The Relevance

The evaluative question quoted below must be included in any process of meta-evaluation:

"In my view, one of the most important questions professional evaluators should regularly consider is the extent to which evaluation has made a contribution to the welfare of humankind and, more generally, to the welfare of the planet we inhabit - and, while we're at it, to the welfare of the other celestial bodies we are beginning to invade" [19].

To discuss procedures of meta-evaluation is to discuss the quality of the evaluative process. Therefore, it is fundamental to consider also the standards that an evaluator should follow, in the light of pre-established criteria of an evaluation of quality. The formative meta-evaluation plays a decisive role in the design of the evaluative process itself in which it is important to pay attention to:

(a) the quality of the instruments for data collection; (b) the quality of the information collected; (c) the choice and adequate application of techniques and models of data treatment; (d) adequate transmission of the results obtained [20] making their reach and limitations clear; and (e) the quality of the evaluators [21].

These are crucial aspects of meta-evaluation since failures in these aspects can often lead to evaluative processes that fail or present results that do not help in the adequate understanding of reality. The execution of a meta-evaluation process must help consolidating and improving knowledge, in the area of evaluation, of those who carry out the evaluative process [22].

With respect to the quality of the instruments for data collection, it is necessary to ensure that they are adequate for obtaining the data one really intends to collect. The validation of the instruments for data collection can be made in different ways. In the case of instruments that collect qualitative information, such validation can be made with the assistance of specialists in the area or through a comparison among different evaluators, techniques, and instruments. In the case of quantitative information, the use of a Confirmatory Factor Analysis is recommended [23]. This is a technique of reduction of data dimension just like Exploratory Factor Analysis, which is better known and more frequently used. The fundamental difference is that Confirmatory Factor Analysis is carried out from the application of a model of structural equation and, therefore, a theoretical model is assumed beforehand relating latent variables (not observable) to the observable variables. In the Exploratory Factor Analysis, on the other hand, each and every latent variable may have an influence on the observable variables, since the number and nature of the latent factors before processing the analysis is unknown. It is precisely this difference that makes the first one confirmatory and the second one eminently exploratory. Another important difference is that in the Confirmatory Factor Analysis the errors are also modelled and may (or may not) be correlated, whereas in Factor Exploratory Analysis it is assumed that the errors may not be correlated (which is not always true). In the Confirmatory Factor Analysis, the model previously established is adjusted for the purpose of minimizing calculated residues through the difference between the variance and covariance matrixes observed and calculated. The Exploratory Factor Analysis is vastly employed, without any type of test for checking whether the errors are not, in fact, correlated. Ideally, different types of instruments for data collection should be used, contemplating preferably quantitative and qualitative information.

With regard to the quality of information, there are two aspects that must be observed: (i) the quality and adequacy of the sources of information and (ii) the adequate treatment of data bases. When choosing the sources of information, it is important to ensure (whenever applicable) that all different groups of possible informants about the evaluative focus are considered. On the other hand, after collecting data and before calculating the indicators, it is fundamental to remove from the data bases information that does not reflect the latent trace one wants to measure. Thus, in the case of instruments that collect quantitative data, those that present responses with always the same pattern, no responses, or indication of objection, must be excluded, as well as any other instruments that when filled in do not reflect a useful information regarding to what one wants to measure [24]. When the information is a qualitative one, this problem may be avoided through data triangulation [25]. In choosing the most adequate technique for modelling and data analysis, it is important to be clear about which evaluative question or questions one intends to answer, since an adequate technique to search for the answer to a given question may not be adequate for another one [26].

7.2.3 Traditional Techniques

A meta-evaluation can be carried out in several ways, through the use of different instruments, but the use of checklists has been the procedure most adopted by many evaluators and evaluation centers, generating satisfactory results [27]. Checklists are defined by the Websters Collegiate Dictionary as "a list of things to be checked or done". In practice, when a checklist is transformed into an instrument for data collection, a set of instruments is created where each item is a statement and the respondent must simply indicate whether the statement is true or not. In the specific case of meta-evaluation, such statement tries to investigate the presence of the patterns adopted in the evaluative process which is the focus of meta-evaluation. Checklists represent an efficient instrument, in a friendly format, for sharing lessons learned in practice [28]. Checklists are very useful because, in the complex world of evaluation, where it is necessary to pay attention to the most varied aspects, it is important to ensure each criterion is being observed.

The Evaluation Center of the Western Michigan University has made available for public use (www.wmich.edu/evalctr) a set of checklists, based on the thirty standards of a true evaluation proposed by the Joint Committee for

Educational Evaluation. It is important to point out that when checklists based on classical logic are used, the result of the meta-evaluation is achieved through a kind of counting the number of true responses. When there are criteria that "do not apply" to a certain evaluation, it is necessary to change the way of doing this counting.

Since meta-evaluation is a type of evaluation, in theory it can make use of different evaluative approaches, different techniques of data collection and different methods of data treatment. However, in the current literature, the use of traditional checklists stands out. This probably occurs because the concept of meta-evaluation is relatively new and its practical applications are recent.

7.3 Fuzzy Logic: Basic Concepts

Concepts of Fuzzy Set Theory and of Fuzzy Logic can be used to translate, in mathematical terms, the imprecise information expressed by a set of linguistic IF-THEN rules. If a human being is capable of articulating its reasoning as a set of IF-THEN rules, it is possible to create an inference system, the algorithm of which may be implemented through a computer program, where Fuzzy Set Theory and Fuzzy Logic provide the mathematical tools for dealing with such linguistic rules [29].

Fuzzy Logic studies the formal principles of approximate reasoning and is based on Fuzzy Set Theory. Fuzzy Logic deals with intrinsic imprecision, associated with the description of the properties of a phenomenon, and not with the imprecision associated with the measurement of the phenomenon itself. While classical logic is of a bivalent nature ("true" or "false", "0" or "1"), fuzzy logic admits multivalence.

In the ordinary set theory, the concept of membership of an element to a set is very precise: either the element belongs or does not belong to a given set. Given a set A, in a universe X, membership of an element x of X to the set A is expressed by the characteristic function A:

$$f_A(x) = \begin{cases} 1 & \text{if } x \in A \\ 0 & \text{if } x \notin A \end{cases}$$

Zadeh generalized the characteristic function so that it could assume an infinite number of values in the interval $[0, 1]$. Given a *fuzzy* set A, in a universe X, membership is expressed by the function $\mu_A(x) : X \to [0, 1]$. A is now represented by a set of ordered pairs $A = \{\mu_A(x)/x\}\ x \in X$, where $\mu_A(x)$ indicates to what extent x is compatible with a set A.

The support set of a *fuzzy* set A is the set of elements in the universe X for which $\mu_A(x) > 0$. Thus, a *fuzzy* set may be seen as the mapping of the support set in the interval $[0, 1]$. If universe X is discrete and finite, the *fuzzy* set A may be represented by a vector with the grades of membership to set

A of the elements corresponding to X through the notation: $\sum_{i=1}^{n} \mu_A(x_i)/x_i$. On the other hand, if the universe X is continuous, the following notation is used: $\int_X \mu_A(x)/x$.

A linguistic variable is a variable whose values are names of fuzzy sets. For example, the answer to an item of a certain instrument of data collection may be a linguistic variable assuming the values "poor", "good", and "excellent". These values are described by means of fuzzy sets, defined by membership functions.

Consider the linguistic variable *scale* (in an instrument for data collection), with values *poor*, *good* and *excellent*, defined by the membership functions shown in Figure 7.1. Answers up to 2.5 have a membership grade equal to 1 in the *poor* set; the membership grade in this set decreases as the response increases. An answer of 5 is considered "totally compatible" with the *good* set, whereas responses above 5 present a membership grade different from zero in *excellent*.

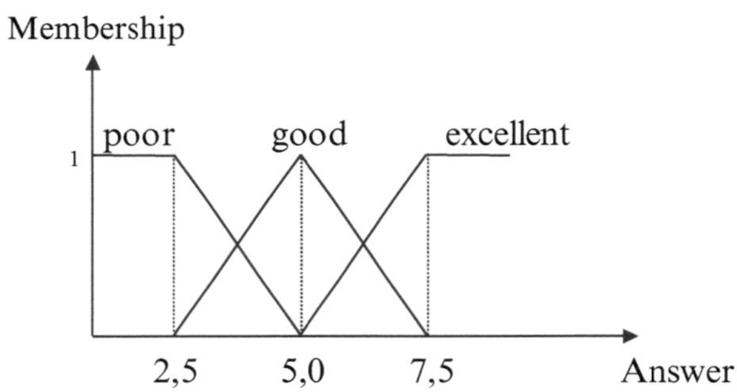

Fig. 7.1. Example of functions of pertinence

Membership functions may have different shapes, depending on the concept that one wishes to represent and the context in which they will be used. Context is extremely important, since the concepts of *poor*, *good* and *excellent*, for example, are extremely subjective. Membership functions may be defined by the user, but they are usually of a standard form (triangular or gaussian, for example). In practice shapes can be adjusted, in accordance with the results, in a trial-and-error procedure.

In fuzzy logic, a conditional statement (IF x is A THEN y is B) is expressed mathematically by a membership function $\mu_{A \to B}(x, y)$, which denotes the *degree of truth* of the *implication* $A \to B$. Some examples of $\mu_{A \to B}(x, y)$, obtained through the simple extension of bivalent characteristic functions of classical logic to fuzzy logic, are:

$$\mu_{A \to B}(x, y) = 1 - \min[\mu_A(x), 1 - \mu_B(y)]$$

$$\mu_{A\rightarrow B}(x,y) = \max[1-\mu_A(x), \mu_B(y)]$$

As to inference, the *modus ponens* of propositional logic (premise 1: x is A; premise 2: IF x is A THEN y is B; consequence: y is B) is extended to the *generalized modus ponens*, described as:

Premise 1: x is $A*$.

Premise 2: IF x is A THEN y is B.

Consequence: y is $B*$

While in classical logic the rule generates a consequence only if premise 1 is the exact antecedent of the rule (and the result is exactly the consequent of that rule), in fuzzy logic a rule is activated if there is a degree of similarity different from zero between premise 1 and the antecedent of the rule. The result will be a consequent with a degree of similarity to the consequent of the rule.

A Fuzzy Inference System, shown in Figure 7.2, does the mapping from precise (crisp) inputs to crisp outputs. The crisp inputs may be measurement or observation data, which is the case of the large majority of practical applications. These inputs are *fuzzified* (mapped to fuzzy sets), which can be viewed the activation of relevant rules for a given situation. Once the output fuzzy set is computed through the process of inference, a *defuzzification* is performed, since in practical applications crisp outputs are generally required. The *rules* are linguistic IF-THEN statements sentences and constitute a key aspect in the performance of a fuzzy inference system.

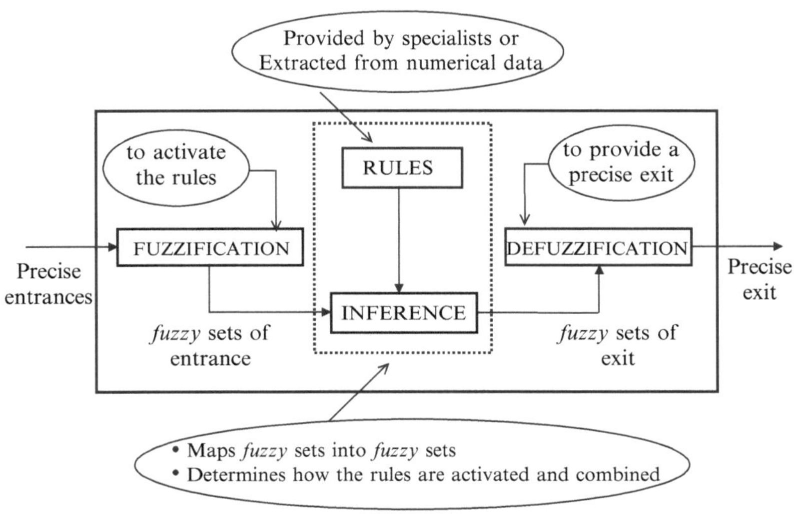

Fig. 7.2. A generic Fuzzy Inference System.

7.4 The proposed methodology

7.4.1 Introduction

The methodology developed in this work (Figure 7.3) is composed of an instrument for data collection (Checklist for the Meta-evaluation of Programs/Projects) and of a fuzzy inference system to treat data related to the meta-evaluation of projects and programs. The instrument for data collection was constructed from the adaptation of the checklist for meta-evaluation developed by the Evaluation Center of the Western Michigan University and the fuzzy inference system was based on the thirty standards developed by the Joint Committee on Standards for Educational Evaluation.

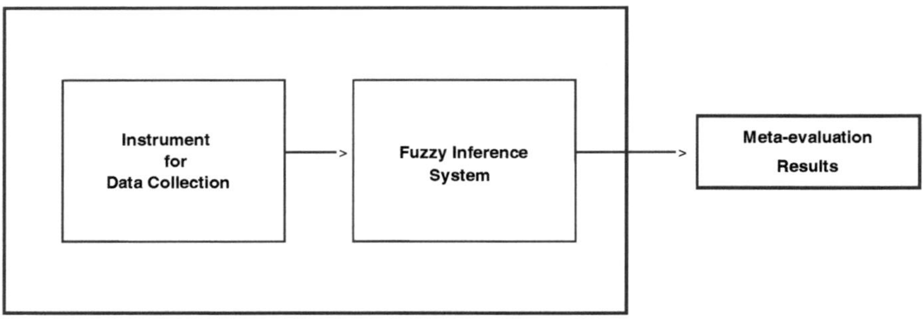

Fig. 7.3. Proposed Methodology of Meta-Evaluation based on a Fuzzy Inference System

Due to the complexity of the problem, the fuzzy inference system was subdivided into thirty-six rule bases, organized into a hierarchical structure composed of three levels (Figure 4):

- Level 1: standards rule bases
- Level 2: categories rule bases
- Level 3: meta-evaluation rule bases

While in the original instrument the respondent could only indicate whether a certain criterion was met or not, the new instrument makes it possible for a respondent to use a scale of eleven points (from 0 to 10) in all the six criteria of the thirty standards. As each standard has six criteria, the instrument has a total of 180 items.

7.4.2 The Instrument

In the development of an instrument for data collection, several aspects need to be taken into account:

(i) quality and validity of the content;
(ii) possibility of accepting intermediary responses between the criterion of excellence considered and the criterion of excellence not considered;
(iii) ease of understanding and of filling out;
(iv) flexibility to be used by different types of evaluators in the meta-evaluation of different evaluative approaches employed in different types of programs or projects; and
(v) possibility of being used by the very team that carries out the evaluation (self-meta-evaluation), by an outside meta-evaluator, or by an independent meta-evaluator.

The quality and the validity of the content were ensured through the creation of the instrument from the adaptation of a checklist, already validated and recognized by those that study and that perform evaluation. However, as it has already been pointed out, the evaluator only indicates whether a given criterion of excellence was met or not, which in many cases may represent a limitation; a partial fulfillment of the criterion is frequently observed. This problem was solved by altering the format of the response in the new instrument the evaluator assigns a grade from zero to ten to the criterion, by using a scale that is already natural to human beings. Grade zero indicates that the criterion was not met; grade ten indicates that the criterion was totally met and the other grades are intermediary answers.

To facilitate the reading and the filling out of the instrument, this was organized in four blocks, one for each category of a true evaluation. General instructions about the use of the document are clearly presented on its first page. A brief explanation of what each category intends to ensure was included in the beginning of each block. The abbreviated concept of each standard is explained before the respective criteria of excellence, always followed by a brief orientation on the filling out of each item. In this way, the instrument may be used for the execution of a self-meta-evaluation, an external meta-evaluation or an independent meta-evaluation.

The division of the instrument into blocks, according to each category, has two additional advantages:

(i) flexibility, since it makes it possible for different evaluators to fill out each block and
(ii) the possibility of carrying out the meta-evaluation of only one of the categories.

7.4.3 The Inference System

The hierarchical inference system, with the proposed three levels, is shown in Figure 7.4. The whole system was implemented in *MatLab*©, by using the *Fuzzy Toolbox*.

With the objective to reduce the possible number of rules, resulting in a more understandable rule base, as well as to provide partial results - for

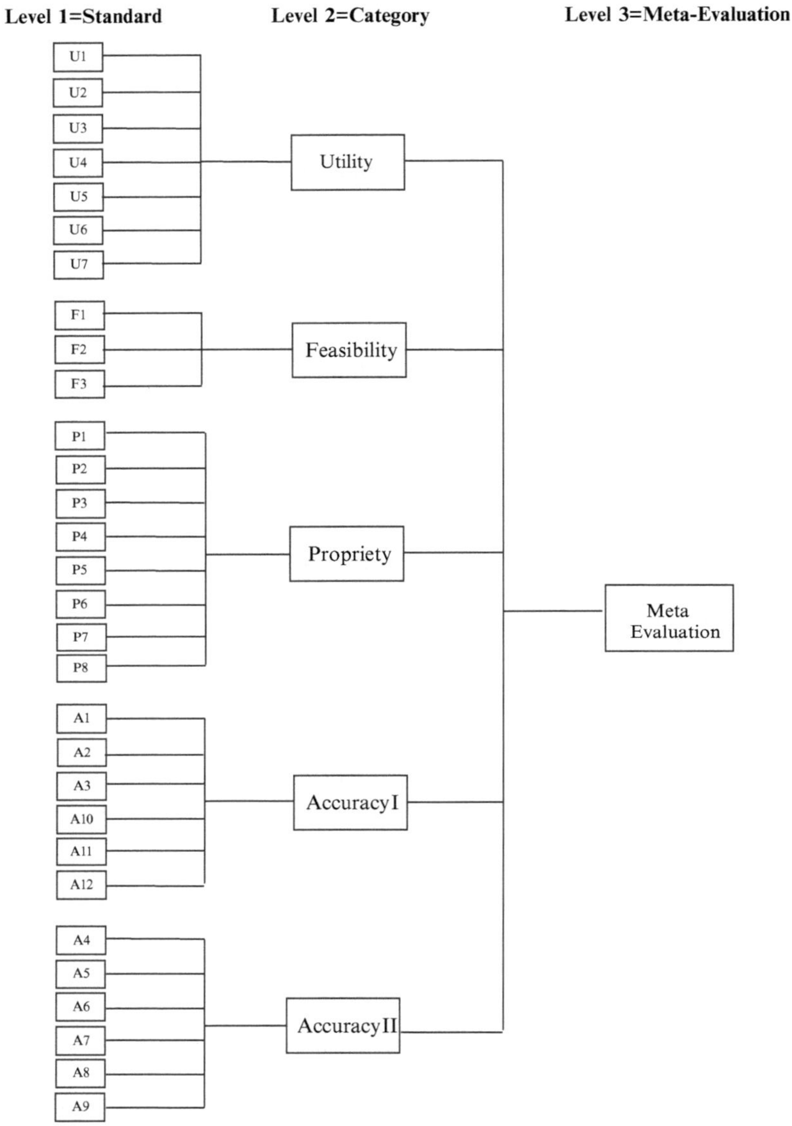

Fig. 7.4. Fuzzy Inference for The Meta-Evaluation of Programs/Projects

standards, categories, and meta-evaluation as a whole - the rule base was built in three levels. Due to the large number of antecedents, the use of a one-level rule base would be impractical: the number of rules would be excessive and the definition of rules by specialists would be almost impossible. The use of a rule base for each category and a set of general rules for meta-evaluation - making use of the results for each of the categories - would also involve an excessive number of antecedents.

The three-level model has thirty six rule bases: 30 in the standards level, 5 in the categories level, and one in the meta-evaluation level. The rule bases were built by considering the guidelines provided by the Joint Committee on Standards and linguistic information provided by qualified specialists.

First, the standards rule bases (level 1) were built. Since each standard is evaluated on the basis of six criteria, the rules at this level have at most six antecedents. Each criterion has three linguistic values (*insufficient*, *satisfactory* and *excellent*) associated with it. The membership functions of there three fuzzy sets are shown in Figure 7.5.

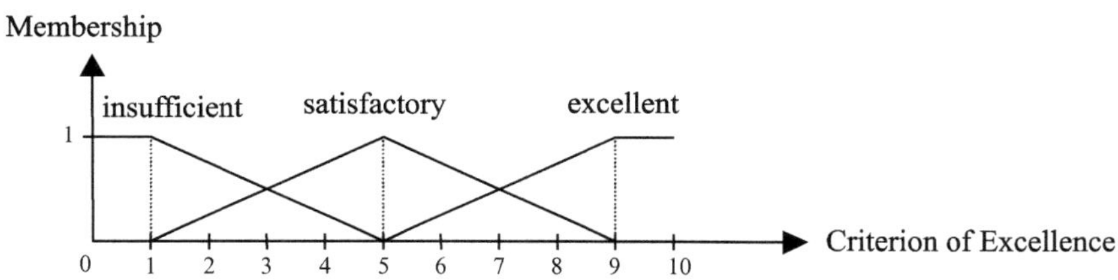

Fig. 7.5. Linguistic values for level 1 input variables

The consequent of each rule is the linguistic variable that represents a specific standard. To this output variable the values *insufficient*, *satisfactory* and *excellent* are also associated, as shown in Figure 7.6. Thirty standards rules were developed.

In the rule bases of level 2, the number of antecedents varies in accordance with the number of standards present in each category, that is: category utility has seven; category feasibility has three; category propriety has eight; and category accuracy has twelve. In the case of the category accuracy, the large number of input variables would jeopardize the development and understanding of linguistic rules. Therefore, the solution was to create two rules bases for accuracy: one with the standards that are directly related to information and its quality (accuracy I rule bases) and another with standards that refer to the analysis and disclosure of information (accuracy II rule bases).

The inputs to the inference systems of level 2 are the outputs from level 1 (linguistic variables with three values each). In the rule bases of level 2, the outputs are the linguistic variables that represent the category and have

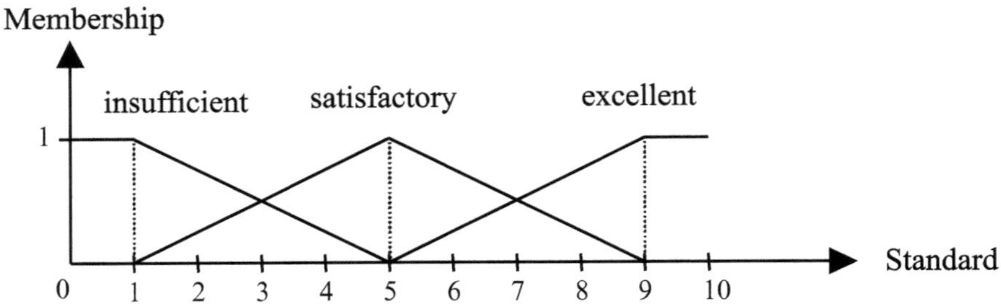

Fig. 7.6. Linguistic values for output variables - Level 1

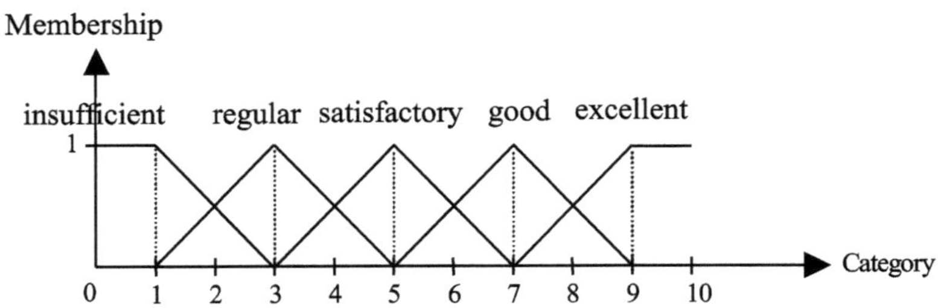

Fig. 7.7. Linguistic values for level 2 output variables

five associated values (*insufficient*, *regular*, *satisfactory*, *good* and *excellent*), specified by fuzzy sets defined by the membership functions shown in Figure 7.7.

Results by category are obtained at level 2 and not only facilitate the elaboration of recommendations and adjustments but also enable some categories to go through the process of meta-evaluation at different moments, when the instrument is used in a formative character.

The meta-evaluation rule base is responsible for the generation of the final result. Rules at this level have five antecedents (outputs from level 2). The consequent is a linguistic variable with five values, as shown in Figure 7.8.

When the instrument is totally filled out, the inference system computes thirty six results. Thus besides calculating a result for each standard, the system also calculates one for utility, one for feasibility, one for propriety, two for accuracy, and the general result of evaluation. Each one of the standards results may be *insufficient*, *satisfactory* or *excellent* and the others: *insufficient*, *regular*, *satisfactory*, *good* or *excellent*.

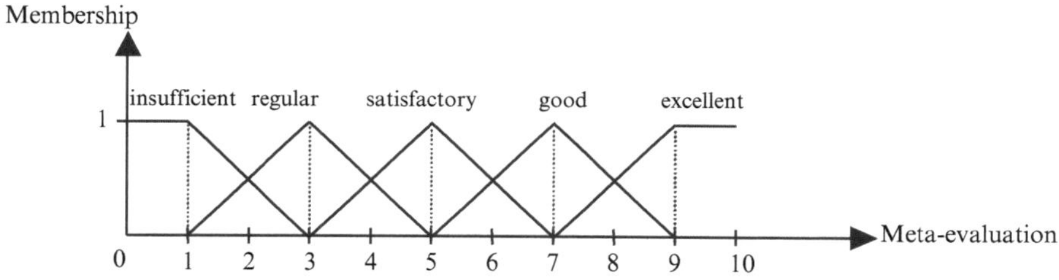

Fig. 7.8. Linguistic values for level 3 output variables

7.5 Case study

To test and validate the proposed methodology, the evaluation of an educational program was considered [30]. The steps followed for data collection and validation of the methodology proposed are shown in Figure 9. In the first step, five types of respondents were selected: EPP - evaluator who participated in the process, ENPP - evaluator who did not participate in the process, PPPE - professionals who participated in the implementation of the evaluated program, META - meta-evaluator who conducted the external meta-evaluation, and STU - student of evaluation who did not participate in the process. In the next step, all the selected respondents received a document with the description of the evaluative process, the list of verification for the meta-evaluation of programs and projects, and a request to assign grades (insufficient, average, satisfactory, good, excellent) to the meta-evaluation in general and to each of the four categories. The META, PPPE and ENPP respondents were also requested to provide an opinion. Respondents could ask for additional clarification (step 3) about the focus of the evaluative process. After the instrument of collection was returned (step 4), the Fuzzy Inference System was fed with data (step 5) for initial processing (step 6). In step 7, discussions about the preliminary results provided by the Fuzzy Inference System were conducted, so that adjustments could be made (step 8). A new processing was then carried out (step 9), new results were provided by the Fuzzy Inference System and, in view of that, the proposed methodology was considered as validated. Finally, a final discussion of the final results was carried out (step 11).

The results provided by the Fuzzy Inference System attained great coherence with the opinions of EPP, ENPP, and meta-evaluators about the evaluation of the educational program. This can be attributed to the fact that those three groups are composed, in general, of professional evaluators with a vast experience in the area of evaluation. Every human being is capable of conducting evaluations, which is generally done several times a day. However, a professional evaluator is one capable of carrying out the evaluation, that is, to produce a value judgment, on the basis of criteria of excellence and external values previously established. This is a competence developed through the study of evaluation and the development of an evaluative culture.

In the case of PPPE, the results are usually coherent, with some exceptions. It should be noted that the professionals who took part in the implementation of the educational program have participated in evaluations and are part of an organization that fosters the evaluative culture. Therefore, those are people in contact with the practice of professional evaluation, which probably justifies the coherence the results.

In the group of students, the results provided by the Fuzzy Inference System showed some discrepancy to their opinion about the evaluation. This can be explained by the fact that students do not have much experience in the area and are still acquiring theoretical knowledge in the field of evaluation. Therefore, it is natural that they may encounter some difficulty to carry out the evaluation and to tell apart their personal opinions, based only in their standards and values, from a value judgment achieved in the light of criteria of excellence.

A more detailed description of the case study as well as the results obtained are presented in [30].

7.6 Discussion

Similarly to what has been observed in evaluative processes, it is difficult to define the correct or the best methodology for carrying out a meta-evaluation. Different methodologies may be better suited to distinct cases. In the proposed methodology it is possible to point out the following advantages:

- With respect to the instrument for data collection, it allows for intermediary answers, which facilitates filling it out, especially in the case of meta-evaluators who still lack a great deal of experience (as is the case in Brazil). The Brazilian evaluator faces diversified demands from a complex context and attempts to fully accomplish the role without the benefit of adequate preparation and working conditions.
- Regarding the use of fuzzy inference systems:
  - i. It utilizes linguistic rules provided by specialists; this favours understanding and the update of rules;
  - ii. It may incorporate contradictory rules, which is not possible when traditional logic is used;
  - iii. It can deal with intrinsic imprecision that exists in complex problems, as is the case of meta-evaluation.
  - iv. It was built on the basis of standards of a true evaluation of the Joint Committee on Standards for Educational Evaluation (1994); thus, it is able to reach a broad range of users.
- With respect to the capacity for adaptation to specific needs:
  - i. Whenever an evaluative process is initiated, it is necessary to verify which are the standards that apply to the evaluation. Frequently, there

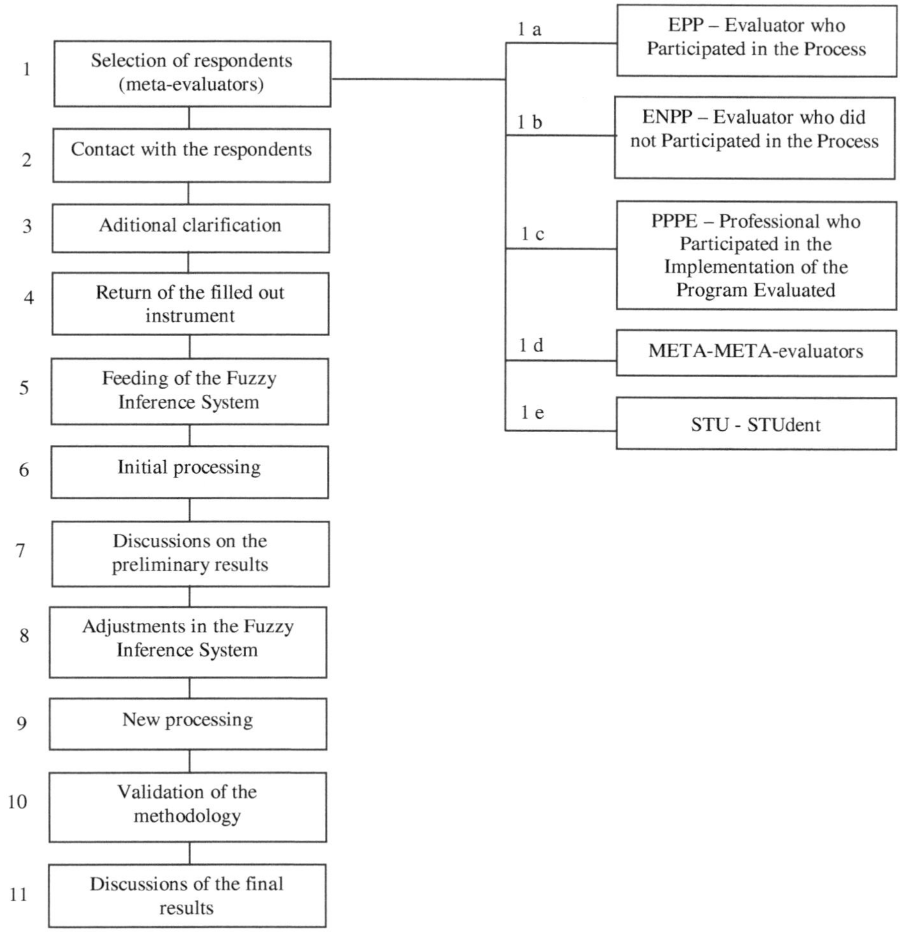

Fig. 7.9. Procedure for Validation of the Proposed Methodology

are cases where one or more of the criteria defined by the Joint Committee are not applicable, which makes it necessary to recalculate the grade attributed to standard and category. With the methodology proposed here, the non-applicable item may be simply not considered (not answered).

ii. The instrument for data collection and the inference system can be adapted to any other set of standards as long as it is possible to translate the knowledge of specialists in terms of IF-THEN rules.

- Regarding transparency, the use of linguistic rules makes it easier to understand and to discuss the whole process.

As for disadvantages, the following could be pointed out:

- an increase in the amount of time spent in filling out the instrument of data collection, as a result of the larger number of options in each item;
- the need to use a computerized system to process the data and calculate the results.

7.7 Conclusion

This paper presented a methodology for meta-evaluation based on fuzzy logic. This new methodology makes use of a Mamdani-type inference system and consists of 37 rule bases organized in three levels: standard (level 1), category (level 2), and meta-evaluation (level 3). The hierarchical structure and the rule bases were built in accordance with the standards of a true evaluation proposed by the Joint Committee on Standards for Educational Evaluation [18]. It is expected that this work may somehow contribute to evaluators, to those who order evaluations, and to those who use the results.

References

1. R. E. Stake. *Standards-Based & Responsive Evaluation.* Sage, Thousand Oaks, CA, USA, 2004.
2. T. Penna Firme, R. Blackburn, e J. V. Putten. *Avaliao de Docentes e do Ensino.* Curso de Especializao em Avaliao Distncia (Organizao Eda C.B. Machado de Souza). V. 5 Braslia, DF.Universidade de Braslia/Unesco, 1998. (in portuguese).
3. M. Scriven, *Evaluation thesaurus* (4th ed.), Sage, Newbury Park, CA, 1991.
4. D. Stufflebeam, "The methodology of meta-evaluation as reflected in meta-evaluations by the Western Michigan University". *Journal of Personal Evaluation*, Evaluation Center, Norwell, MA, v. 14, n. 1, 2001, p. 95-125.
5. G. J. Klir & B. Yuan, *Fuzzy Sets and Fuzzy Logic* - Theory and Applications, Prentice Hall PTR, 1995.
6. M. Kochen. Aplication of Fuzzy sets in psychology. In: Zadeh, L.A. et. Al. (Editor). *Fuzzy sets and their applications to cognitive and decision process.* Academic Press, USA, New York, 1975, pp. 395-407.
7. E. H. Mamdani, S. Assilian. "An Experiment in Linguistic Synthesis with a Fuzzy Logic Controller". *International Journal of Man-Machine Studies*, 7(1), 1975, pp.1-13.
8. M. Scriven. The methodology of evaluation. In R. E. Stake (editor) *Perspectives of Curriculum Evaluation* (American Educational Research Association Monograph Series on Curriculum Evaluation, N. 1), Rand McNally.Chicago, IL, USA, pp.39-83, 1967.
9. Patton, M. Q. (1997). *Utilization-Focused Evaluation* (2nd ed.). Sage, Thousand Oaks, CA, USA, 1996.
10. M. Q. Patton. The roots of utilization focused evaluation. In: M. C. Alkin (ed), Evaluation Roots - *Tracing Theorists' Views and Influences.* Sage Thousand Oaks, 2004, pp. 276-292.
11. R. E. Stake. *Standards-Based & Responsive Evaluation*, Sage, Thousand Oaks, CA, USA, 2004.

12. R. E. Stake, Stake and Responsive Evaluation. In: M. C. Alkin (editor), *Evaluation Roots Tracing Theorists' Views and Influences*. Sage, Thousand Oaks, 2004, pp. 203-276.
13. E. W. Eisner. *The educational imagination: on the design and evaluation of educational programs* (3rd ed). Macmillan, New York, New York, 1994.
14. E. W. Eisner. *The enlighted eye*. Macmillan, New York, New York, USA, 1991.
15. W. R. Shadish, T. D. Cook, L. C. Leniton. *Foundations of Program Evaluation: Theories of Practice*. Newbury Park, CA: Sage, 1991.
16. Joint Committee on Standards for Educational Evaluation. *Standards for Evaluations of Educational Programs, Projects and Materials*. Sage Publications USA, 1981.
17. Joint Committee on Standards for Educational Evaluation. *The Personnel Evaluations Standards*. Sage, Newbury Park, CA, USA. 1988.
18. Joint Committee on Standards for Educational Evaluation. *The Program Evaluation Standards*. 2nd ed, Sage Publications, USA, 1994.
19. M. Scriven. Reflections. In: Alkin, M. C. (Editor) *Evaluation Roots: Tracing Theorists' Views and Influences*. Sage Publications. Thousand Oaks, USA, 2004, p.p. 183-195.
20. B. A. Bichelmeyer. *Usability Evaluation Report*. Western Michigan University. Kalamazoo, USA, 2002.
21. W. R. Shadish, D. Newman, M. A., Scheirer, C. Wye, Cris. *Guiding Principles for Evaluators* (New Directions for Program Evaluation, no. 66). Jossey-Bass. San Fracisco, USA, 1995.
22. C. A. Serpa, T. Penna Firme, A. C Letichevsky. Ethical issues of Evaluation Practice within the Brazilian Political context. *Ensaio: avaliao e polticas pblicas em educao*, revista da Fundao Cesgranrio, Rio de Janeiro, v. 13, n. 46. 2005.
23. K. A. Bollen. *Structural Equations with Latent Variable*. John Wiley & Sons. New York, USA. 1998.
24. A. Bryk, S, Raudenbush, S. *Hierarchical Linear Models*. Sage Publications, Newbury Park, CA, USA, 1992.
25. P. H. Franses, I. Geluk V. P. Homelen. *Modeling item nonresponse in questionnaires*. Quality & Quantity, vol.33 1999, pp. 203-213.
26. A.C. Letichevsky. *La Categoria Precisin en la Evaluacin y en la Meta-Evaluacin: Aspectos Prcticos y Tericos*. Trabajo presentado en la I Conferencia de RELAC. Lima. Peru. 2004 (in spanish).
27. T. Penna Firme, A. C. Letichevsky. O Desenvolvimento da capacidade de avaliao no sculo XXI: enfrentando o desafio atravs da meta-avaliao. *Ensaio: avaliao e polticas pblicas em educao*, revista da Fundao Cesgranrio, Rio de Janeiro, v. 10, n. 36, jul./set. 2002 (in portuguese).
28. D. Stufflebeam. *Empowerment Evaluation, Objectivist Evaluation, and Evaluation Standards: Where the Future of Evaluation Should not go and where it needs to go*. Evaluation Practice.Beverly Hills, CA, USA. 1994, pp. 321-339.
29. J. M. Mendel, *Fuzzy Logic Systems for Engineering: a Tutorial*, Proc. IEEE, V. 83, No. 3, 1995, pp. 345-377.
30. A.C.Letichevsky, *Utilizao da Lgica Fuzzy na Meta-Avaliao: Uma Abordagem Alternativa*, PhD Thesis, Department of Electrical Engineering, PUC-Rio, 20 de fevereiro de 2006 (in portuguese)

8

Evaluation of Fuzzy Productivity of Graduate Courses

Annibal Parracho Sant'Anna

Departamento de Engenharia de Produo - Escola de Engenharia Universidade Federal Fluminense Rua Passo da Ptria, 156 24210-240 Niteri-RJ
Fax:5521-26295435 tppaps@vm.uff.br

The data set for this application is provided by official reports delivered by the Master Programs in Production Engineering in Brazil. A set of fuzzy criteria is applied to evaluate their Faculty productivity.

The fuzzy approach taken allows considering, besides the goals of maximizing output/input ratios, more asymmetric criteria. For instance, productivity may be measured, in the orientation to maximize the output, by the probability of the production unit presenting the maximum volume in some output and not presenting the maximum volume in any input. Fuzzy Malmquist indices are derived from these measures to evaluate evolution through time.

These different criteria are applied to annual data on two outputs, number of students concluding courses and of Faculty members with research results published. The results obtained support the hypothesis that time periods longer than one year are necessary to avoid false alarms.

8.1 Introduction

Brazil has an influential system of evaluation of the graduate courses. The system is managed by CAPES - governmental funding agency for higher education that applies the results of the evaluation to support its decisions relative to scholarships and financial support to projects. An institution is allowed to start offering Ph. D. Programs only if its Master Program in the area is graded above good. Courses with low grades are forced into not accepting new students. Other funding organizations also consult CAPES classification, so that, briefly speaking, this evaluation provides a reference for the whole community of research and higher education in Brazil.

CAPES evaluation system is based on the annual sampling of data, automatically summarized in a large set of numerical indicators. The final evaluations are presented in a scale from 1 to 5 for the programs offering only

Annibal Parracho Sant'Anna: *Evaluation of Fuzzy Productivity of Graduate Courses*, Studies in Computational Intelligence (SCI) **44**, 167–182 (2007)
www.springerlink.com

Master Courses and from 1 to 7 in the case of Ph. D. Programs. Such degrees are valid for three years, but annually partial indicators and comments on them are officially published.

The system was developed in close connection with the academic community and strongly influenced by the principle of peers' appraisal. In such a way that, although final decisions on resources allocation and the final grades of the courses must be approved by a Scientific Committee, the power of judgment is concentrated on small committees representing the researchers in relatively small areas of knowledge. These committees assemble, as a rule, twice a year: first to review their criteria and weights and later to examine the information gathered about the courses and issue their evaluations.

The committees are constituted through the appointment by CAPES Board of an area representative for a three years term. The remaining members of the committee are chosen by this area representative and dismissible *ad nutum*. The practical meaning of this appointment system is that the area representatives are the people who must concur with the management ability, define the area goals and understand how they fit in the government global strategy to improve higher education in the country. The researchers they bring to the committee contribute to the evaluation with the appraisal of specific production issues.

This structure has been strong enough to hold through decades. Occasional criticism from experts in evaluation who would like to see decisions more clearly related to a management philosophy, objective goals and well defined liabilities, and from courses managers who do not see their real concerns and achievements effectively taken into evaluation have been easily overridden by the ability of choosing a team of area representatives who correctly mirror the political power balance between the Higher Education institutions.

Nevertheless, the ambiguity involved in the roles played by the members of the committees may, in fact, sometimes distort the evaluation. One example of such a situation is that of the Master Programs in Production Engineering. These courses are evaluated together with those of Mechanical Engineering and a few other areas of expertise more related to Mechanics. Although, in terms of number of courses and students, Production Engineering is in the majority, their area representative has always been more linked to the field of Mechanical Engineering. An explanation may be found in the fact that the Production Engineering Master Programs serve a large number of part-time students who do not consider entering immediately a Ph. D. course, unlike the other Master courses in Science and Engineering, Mechanical Engineering courses included among these. Thus, although possibly more focused on programs management subjects, an area representative coming from the Production Engineering side would represent values, with respect to emphasis on course structure and speed of the formation, in disagreement with the mainstream.

This study intends to bring elements to the discussion of productivity measurement issues specially important in the evaluation of the Master courses.

A main concern of CAPES evaluation system is on not gauging results without taking into account the volume of resources applied. Thus, most of its indicators are built dividing results observed by the number of teachers in the course. Evaluation becomes then based on Faculty productivity. The denominator in the Engineering area, to be more precise, has been until recently the number of people applying at least 30 percent of their work time in the course and present in no less than two of three kinds of activity considered: research, advising and classes teaching. This quantity, named NRD6 (reference core number 6, because there were other 5 slightly distinct criteria applied in other areas), may change considerably from one year to the other without any real effect on the courses results, mainly on those affecting the quality and quantity of students following or concluding the Master courses.

Related to the instability above mentioned, the periodicity of the evaluations is another subject of investigation. Courses coordinators present annual reports to CAPES. Following data with a periodicity smaller than one year generally does not make sense, because presentation of dissertations is irregularly spread though the year and short-term statistics about research papers submission, approval and publication, if the Faculty size is not very large, will also be very irregular. For this reason, policies taken to correct deficiencies detected by the end of the year will make the yearly data present some negative serial correlations. The knowledge of that has made CAPES initially provide evaluations only biennially. Later, however, the periodicity was changed, with grades being made public every three years, but evaluations, named qualitative, sent to the courses every year. With this change, the area representatives felt the obligation to produce evaluations in terms of 'you've improved' or 'you're worse'. Sentences like those may have a strong effect on courses administration. A version of Malmquist index is employed here to investigate the variations of productivity measurements, trying to check the reliability of such annual evaluations.

Malmquist indices constitute then another topic in this article. An entirely new form of derivation of these indices, taking as starting point fuzzy productivity measures, is proposed. The application of this approach leads to clear results. In the case of the courses evaluated, there is high instability in the annual indices. This instability is strongly reduced as we take a period of two years to compute the productivities.

In the next section, the fuzzy productivity measures are defined. Sections 8.3 and 8.4 discuss key concepts of randomness and dependence and determine the procedures applied to completely quantify these productivity measures. Section 8.5 develops the indices to assess evolution chronologically. Finally, in Section 8.6, a comparison of global productivity scores for different years and of Malmquist indices generated from these scores is presented.

8.2 Productivity Evaluation

Different approaches have been developed through time to compare production units according to their efficiency in extracting the largest possible aggregate of products from the smallest aggregate of employed resources. Data Envelopment Analysis (DEA) is an attractive instrument to reach this goal. It measures efficiency in a realistic way by the distance to the best observed performance and, to take into account that each unit may address a proper market niche, aggregation is performed using for each unit evaluated the most favorable weights.

The typical DEA efficiency evaluation algorithm was developed by [1]. The concept of efficiency applied comes from [2]. Difficulties in the interpretation of the results obtained in some practical situations induced the development of alternative algorithms. The first has to do with the scales of operation of the units put in parallel. There are situations in which operation units face orders with a volume determined out of its decision scope, in such a way that they cannot change their size. Their efforts to elevate productivity are driven to minimize the volume of employed resources. Conversely, situations may occur in which the available resources are out of the range of decision of the production unit. The productivity is then given by the volume of the resulting production. The inclusion in the analysis of a unit with dimensions very different from those found in the rest of the group and which assure it advantages of scale may result in this unit receiving a paradigmatic position that is, in fact, not reasonable to expect to be reached by the other units. Algorithms were developed in [3] to deal with returns to scale. A problem with this approach is that any unit with an extreme value becomes necessarily evaluated as fully efficient.

The second difficulty has to do with the freedom of specializing. In the context of multiple dimensions, if it is convenient, any unit is allowed to assign, in the composition of the criterion of efficiency that is applied to it, a null weight to any variable, in such a manner that its performance in relation to that variable does not affect its productivity measure. An alternative algorithm was developed in [4] to force the inclusion in the comparison of the amounts of products and resources disregarded in the direct application of Farrell's approach. This algorithm will measure the efficiency through the sum of the slacks appearing in the optimizing solution instead of the productivity ratio. The onus involved in this is the loss of independence of scale of measurement.

These aspects do not bound the applicability of the methodology since, to correct any misinterpretation, it is enough the careful examination of the scales of production and scales of measurement. A third difficulty comes from the possibility of random errors in the measures of inputs and outputs. These errors may come from imprecise measurement tools or from conceptual distances between the true inputs and outputs and the variables effectively measured to represent them. This difficulty is more serious than the preceding ones, because the distortions that random disturbances may cause are not squealed by any external sign. The excellence frontiers generated by performances reflecting the effect of large measurements errors cannot be distinguished from those generated by observations accurately measured. Alternatives have been developed to deal with this difficulty that depend on being able to parametrically model the frontier ([5]) or to statistically model the efficiencies vector ([6]). [7] has overcome this difficulty by taking a simulation approach. [8] took another way to avoid the efficiency ratio modeling, by separately considering the probabilities of maximizing outputs and of minimizing inputs. The global efficiency criteria developed here follow this last approach. The errors in the initial measurements are modeled with mild assumptions that will affect in a balanced way the computations of the distances of the different units to the frontier.

There are two basic ideas governing this approach. The first is to measure the distance to the frontier according to each input or output in terms of probability of reaching the frontier. This preserves DEA advantages of relating the efficiency to observed frontiers and not being influenced by scales of measurement.

The second basic idea is to take into account all variables and all compared units in the evaluation of each unit, thus softening the influence of extreme observed values. While the frontier of excellence tends to be formed by rare performances, the comparison with a large set of observations with more frequent values makes the evaluation process more resistant to random errors.

An advantage of this fuzzy approach is an automatic reduction of the chance of the very small and the very large units appearing as efficiency benchmarks to units operating on much different scales. This happens because units with extreme values will have their efficiency measured through the product of probabilities very close to zero by probabilities very close to one, while the units with values in the central section will have their measures of efficiency calculated through the product of more homogeneous factors.

Besides centering attention on optimizing inputs or outputs, this approach allows us also to choose between an optimistic and a pessimistic point of view. Optimistically, one would consider enough the optimization of no more than one among the inputs or outputs while, pessimistically, one would evaluate in terms of probabilities of optimizing all inputs or outputs.

Another possible choice is between the frontiers of best and worst values. [9] builds efficiency intervals by considering distances to both frontiers. Let us say that a measure is conservative when it values distance to the frontier of worst performances and progressive when it is based on proximity to the excellence frontier. When considering the inputs, the progressive orientation would measure probabilities of minimizing observed values while the conservative orientation would measure probabilities of avoiding maximum values. Conversely, with respect to outputs, the conservative orientation would consider only probabilities of not minimizing while the progressive orientation would look for the probabilities of reaching the frontier of maximum values.

From these two points of view, the classical DEA efficiency measures are optimistic and progressive. In some instances, to induce innovation without allowing for waste, it might be more fruitful to follow an orientation progressive and optimistic with respect to outputs but conservative and pessimistic with respect to inputs. The efficiency would be, in this case, measured by the probability of maximizing at least one of the outputs while not maximizing any input. A few other efficiency measures are proposed below, built by combining these orientations in different manners.

Other fuzzy measures even less related to the basic DEA criteria may also be considered. They comprise the situation where the variables are not listed as resources and products. One of them is the probability of presenting the best performance according to no more than one criterion, for instance by the maximization of one output or by the minimization of one input. Another alternative is better applicable to the case of two or more independent blocks of criteria. In this case, a global measure might be given by the probability of presenting the best performance in at least one criterion of each block.

The choice of the composition eventually decided will have to take into account practical considerations. There are situations, as in many segments of public administration, where the reference is placed in the worst possible performance. From such worst levels progress starts. In these cases, most observed values and the most reliable ones are near the minimum efficiency level allowed. With only a few sparse values in the efficient extreme, it becomes safer to evaluate the performance by the probability of staying away from the inefficiency border than by the proximity to the efficiency frontier. In fact, in such a situation, any value registered by mistake near the value of a unit that is, in truth, the most efficient would strongly affect the probability of that best unit reaching the efficiency frontier while it will only slightly affect its probability of reaching the inefficiency border.

The most optimistic way to compose criteria attributing equal importance to all of them will evaluate each unit by its probability of being preferred in at least one criterion. In the case of production units employing inputs to produce outputs, the measure of efficiency should then correspond to the probability of maximizing the produced volume of some output or minimizing the employed volume of some input. Denoting by P_{ik} the probability of the k-th unit being in the frontier according to the i-th criterion, if the criteria are independent, the final measure of efficiency of this unit would be $1-\prod(1 - P_{ik})$, for i varying along all considered criteria.

Nevertheless, by placing side-by-side resources and products, this measure does not properly evaluate productivity. According to it, the production units may lift its evaluation by raising the volume of any output offered or reducing the volume of any input employed. It is enough for the unit to present large production of one output or small consumption of one input to reach the maximum efficiency measurement. Low input/output ratios would result in high efficiency evaluations, but the opposite would not necessarily be true.

To avoid rewarding the effort on only increasing the production or, on the contrary, the effort on only reducing consumption, it is enough to require simultaneously high probability of maximization of some output and high probability of minimization of some input. Assuming independence, the probability that the k-th production unit presents the maximum value of some output in the group is given by 1-$\prod(1 - P_{jk})$, with j varying only along the outputs and P_{jk} representing the probability of this production unit presenting the largest value for the j-th output. Analogously, the probability that the k-th production unit presents the minimum value of some input in the group is given by 1-$\prod(1 - P_{ik})$, with i varying only along the inputs and P_{ik} denoting the probability of this unit presenting the smallest value in the sample for the i-th input. In this context, the productivity will then be measured by $[1\text{-}\prod(1 - P_{ik})][1\text{-}\prod(1 - P_{jk})]$ where the first product has one factor for each resource, represented by the index i, and the second product has one factor for each product, represented by the index j. This measure, optimistic and progressive with respect to both sets of variables, will be denoted by $OGOG$, with O for optimistic and G for progressive.

There are situations where the optimistic approach in $OGOG$ must be replaced. In such situations, it is important not being wasteful in the use of any resource and not being negligent in attending the demand for any product. The efficiency could then be measured by the probability of presenting the largest value of each output and the smallest value of each input. Changing the O of optimistic by an E of pessimistic, this measure will be denoted by $EGEG$. It will be given, in the framework of last paragraph, by the product $\prod P_{ik} \prod P_{jk}$.

A problem with $EGEG$ is that, if just one of the best observed performances is the result of a large error, the other units will have not only distorted, but also small, final evaluations. A measure more resistant to the influence of deviations inherent in small measurements is obtained using the probability of not being the worst in any variable, that means, the probability of reaching the highest value in no input and reaching the lowest value in no output. Since it is conservative with respect to inputs and outputs, this pessimistic measure will be denoted $ECEC$. Assuming independence, it will be given by $\prod(1 - P_{jk})$, where j is the index of the product and varies along all variables, those representing inputs and those representing outputs, and P_{jk} denotes the probability of the k-th unit presenting the highest observed value in the j-th variable if this represents an input or the lowest value if it represents an output.

Asymmetric measures, from the optimistic or from the progressive point of view, may also be of interest. If we take an approach conservative and pessimistic to inputs and progressive and optimistic to outputs, the fuzzy measure of efficiency will be given by $\prod(l - Q_{ik})[1\text{-}\prod(l - P_{jk})]$, where the second product is as in the previous paragraph, but in the first one appears

Q_{ik} representing the probability of the employed volume of the i-th resource being the largest in the whole set. This measure will be denoted by *ECOG*.

The reverse measure will be progressive and optimistic when dealing with inputs and treat conservative and pessimistically the outputs. It will be given, analogously, by an expression with two products: $[1\text{-}\prod(1 - P_{ik})]\prod(1 - Q_{jk})$, where now it is in the second product that there are new factors Q_{jk}, denoting the probabilities of the generated volumes of outputs being the smallest. This measure will be denoted by *OGEC*.

Other measures may also be useful. For two inputs and two outputs, there are 16 possible arrangements. If there is only one input or output, the optimistic versus pessimistic distinction does not apply and the number of different measures falls by one half.

Fuzzy measures may also be derived from the productivity ratios. For instance, the probability of maximizing some output/input ratio, or the probability of not minimizing any such ratio. Since the extreme ratios are easily far away from the rest, these measures will present larger instability than those considering inputs and outputs *per se*.

As probabilities, all these absolute measures vary from one to zero. But due to their diverse degrees of exactness, when comparing evaluations according to more than one of them, it is interesting to standardize. This can be done by means of relative measures obtained by dividing the probability corresponding to each production unit by the maximum observed value for such probability in the whole set of units under evaluation.

8.3 Randomization of inputs and outputs

With the introduction of random measurement errors, the volumes of inputs and outputs presented at the beginning as deterministic can be treated as estimates of the means of probability distributions. Estimates of other parameters of these distributions can be derived from the same set of observations, although it is difficult to have a number of observations large enough to estimate higher moments with precision, in the first applications.

In order to compensate for the lack of empirical information, simplifying and equalizing assumptions are the essence of Fuzzy Sets Theory ([10]). In the present situation, imposing assumptions of independence between the disturbances affecting the different observations and of normality with identical variance, will provide an initial framework.

What variance will be large enough to allow for inversion of all ranks with nonnegligible probabilities and not disregard the location estimates observed? Establishing how small the probability of inversion of ranks between the units with the highest and the lowest observed values would complete the statistical modeling under the normality assumption.

We may follow the usual practice of deriving estimates for the dispersion parameter of the disturbance of each measure from a measure of the dispersion in the observed sample. The sample in this case is formed by the observed values in the whole set of examined production units. Given the small number of production units usually observed, to follow again the common practice in quality control, we may use the sample range and derive an estimate for the standard deviation dividing it by the relative range ([11]).

This procedure may be set more formally. Denoting by $d_2(n)$ the normal relative range for samples of size n, if the vector of observations for the i-th variable is $(y_{i1}, \dots, y_{in})$, the whole randomization procedure consists in assuming, for all i and k, that the distribution of the i-th variable on the k-th observation unit is normal with expected value given by y_{ik} and standard deviation given by $(\max y_{i1}, \dots, y_{in} - \min y_{i1}, \dots, y_{in})/d_2(n)$.

It is also possible to abandon the hypothesis of identical dispersion and to increase or reduce the standard deviation of one or another measure to mirror a stronger or weaker certainty regarding the measures provided about better or worse known production units. Nevertheless, dispersion variations are in general difficult to quantify.

The independence between random errors on the measurements of the same input or output in different production units is also a simplifying assumption. If the units are only ranked by pairwise comparison, it may be more reasonable to assume a negative correlation. To model that precisely, it would be enough to assume identical correlations and to derive this identical value from the fact that the sum of the ranks is a constant. This correlation would, however, quickly approach zero as the number of units grows.

Another aspect to be considered when modeling the dispersion is that an identical standard deviation implies a larger coefficient of variation for the measures of smaller values. Therefore, assuming identical distribution for the disturbances implies, in fact, attributing proportionally less dispersion to the larger measures of input and output. In order to imply that the most important measures are taken more carefully by making the smallest dispersions correspond to the values closest to the frontier of excellence, we may work with inverse inputs and then transform in maximization of the inverted input the goal of input minimization. This idea of inversion of opposite values is present in the output/input ratio of DEA models.

8.4 Application of Correlation Estimates

Independence between the criteria is generally desired, to avoid collecting duplicate information and to simplify the interpretation of the results. But, although independent inputs may be chosen, disturbances affecting the outputs should not necessarily be considered independent of those affecting the inputs applied to produce such outputs and, consequently, independent among themselves. One advantage of the fuzzy approach is that estimates of the correlation can be employed to improve the calculation of the joint probabilities.

Estimation of the probability of jointly optimizing two variables can be based on the knowledge of the correlation coefficient and of the marginal probabilities. The correlation coefficient between two Bernoulli random variables is obtained, by definition, dividing the difference between the probability of joint choice and the product of the marginal probabilities by the square root of the product of four terms, these marginal probabilities and their complements. From this we derive an estimate for the joint probability as the sum of two parcels: the first parcel is the product of the marginal probabilities and the second is an estimate of the correlation coefficient multiplied by the square root of the product of these marginal probabilities of optimizing and not optimizing each of the two variables.

To estimate the correlation coefficient between the indicators of optimization of the two variables, we start with the correlation coefficient between the vectors of probabilities of the different units optimizing each variable. This sample correlation coefficient may overestimate the absolute value of the correlation, leading eventually to negative or larger than one estimates for the joint probabilities. To avoid that, we move from this initial estimate towards the independence assumption, by bounding the joint probability, from above, by the minimum of the marginal probabilities and, from below, by one half of the product of these marginal probabilities. This limit of one half reflects the idea that the values observed at the unit should receive at least the same importance given to evidence extracted from the whole population. Other values may be tried, taking the result under independence as a reference to simplify comparisons.

Formally, the estimate of the probability of the k-th unit presenting jointly the maximum value in the i-th and j-th variables, given the marginal probabilities P_{ik} and P_{jk}, according to the procedure above outlined will be initially given by $P_{ijk} = P_{ik}P_{jk} + r_{ij}/\sqrt{P_{ik}P_{jk}(1-P_{ik})(1-P_{jk})}$, where r_{ij} is the correlation coefficient between the vectors $(P_{i1}, \dots , P_{in})$ and $(P_{j1}, \dots, P_{jn})$ of all probabilities of presenting the maximum value in the i-th and j-th variables. If this sample correlation coefficient is such that, for some k, P_{ijk} is larger than P_{ik} or P_{jk} or smaller than $P_{ik}P_{jk}/2$, it is replaced by the nearest value satisfying these constraints.

To compute the joint probabilities of more than two events, what will be frequently necessary, we may apply sequentially the procedure above sketched. To reduce the diffusion of rounding errors we should start with the variables that are known to be uncorrelated. If there are not two such variables, since the inputs are more likely to present low correlations, we shall start with the two inputs presenting the smallest sample correlation coefficient. If there is only one input, it must be correlated with all outputs; then, we shall start with the two outputs with the smallest sample correlation coefficient. After that, sequentially, enter the variables with the vector of marginal probabilities less correlated to the last vector of joint probabilities obtained.

8.5 DEA and Malmquist Indices

The Malmquist productivity index was introduced by [12], following [13] measure of evolution of productivity set in terms of distance to the best observed productivities. [13] index is a quantity index not depending on revenues or cost shares to aggregate outputs or inputs. The independence of weights or shares is also a characteristic of DEA. Exploring that, [14] proposed a Malmquist index based on distances to the DEA efficiency frontier.

To consider the shifts on the frontier, the most employed DEA-Malmquist indices are the geometric mean of two indices, one relative to the initial frontier and the other relative to the subsequent frontier. The first is calculated dividing by the initial efficiency of the unit under evaluation the efficiency of a hypothetical unit introduced in the initial data set in the place of such unit and with the inputs and outputs observed in it in the subsequent year. In the same way, the second is obtained dividing the efficiency of the unit in the second year by that obtained substituting for their input and output values those of the previous year.

The same idea is applied here, using, instead of DEA distances to the frontier, the fuzzy productivity measures. For instance, generally denoting by $OCEG(k,s,t)$ the OCEG measure of global productivity of the k-th unit when its inputs and outputs of year s are introduced in the computation together with the inputs and outputs observed on year t in the other units, the $OCEG$-Malmquist index for the k-th unit from year t to year $t+1$ will be given by the square root of the product of $OCEG(k,t+1,t)/OCEG(k,t,t)$ by $OCEG(k,t+1,t+1)/OCEG(k,t,t+1)$.

In addition to comparing the values associated with the different fuzzy measures, we may also pursue the decomposition of the indices in components measuring efficiency change and technical change. As in the framework of [14], the index of efficiency change will be the ratio between the productivity evaluations of the unit in the two years and the technical change will be given by the ratio between the geometric mean Malmquist index and the index of efficiency change.

8.6 Evaluation of Productivity in Production Engineering Courses

This section applies the fuzzy productivity measures and Malmquist indices above discussed to analyze the Master Courses in Production Engineering in Brazil during the period from 1998 to 2001. Alternative approaches to the measurement of productivity in the University are discussed in [15] and [16].

One single input was studied, the Faculty number, measured by the NRD6 CAPES concept. In the output side, two products are examined: number of Faculty members presenting new research reports and number of students presenting final dissertations. These three variables are proxies to difficult to measure concepts: the volume of academic resources employed, the knowledge produced and the human resources qualification generated. The distance between these basic concepts and the measurements obtained is theoretically filled by the random measurement errors. To evaluate the effect of imprecision in these measurements, together with the analysis being based on yearly data, another one was made, with the average input and outputs values for the initial and the final pairs of years.

Table 8.1 presents the entrance data, for 2000 and 2001. Table 8.2 and Table 8.3 present the global productivities for these two last years separately and together, according to four aggregation concepts: *OGOG*, *OGEG*, *OCEG* and *OGEC*, assuming normality, independence between disturbances affecting measurements in different observation units and estimating standard deviations and correlations between outputs and between aggregate output and input as proposed in Sections 3 and 4. It is interesting to contrast these two years because 2000 was the third of a CAPES three years evaluation cycle and 2001 the subsequent year.

Table 8.1. Inputs and Outputs

	Permanent Faculty		Faculty Publishing		Dissertations	
Course	2000	2001	2000	2001	2000	2001
Sc	103		80		277	
Rj	28	23	16	21	77	67
Sp	25	17	19	16	22	22
Fscar	21	24	19	24	25	27
Sm	21	27	10	10	18	48
Espb	20	11	15	10	5	12
Ff	16	20	11	12	22	19
Cefet	11	15	11	13	14	22
Pe	10	10	9	9	16	14
Mep	11	13	11	12	13	15
Spscar	10	11	8	9	22	26
Pb	9	6	8	5	13	18
Rgs	8	9	7	8	22	12
Puc	10	10	8	8	27	19
Mg	9	5	9	3	16	10
P	8	7	8	7	22	22
Ei	8	9	6	7	11	12
Mean	19.3	13.6	15.0	10.9	36.6	22.8
St. Dev.	23.1	6.9	17.6	5.6	65.6	15.1

Comparing the measures, in 2000, the main differences are between *OCEG* and the other measures. Treating the input conservatively, *OCEG* favors Sc because its probability of escaping the frontier of largest input is much large than its probability of reaching the frontier of lowest input. The withdraw

Table 8.2. Fuzzy Global Productivities for 2000 and 2001 separately

	2000				2001			
Course	OGOG	OGEG	OCEG	OGEC	OGOG	OGEG	OCEG	OGEC
Sc	1%	2%	100%	0%				
Rj	83%	77%	12%	33%	34%	33%	8%	10%
Sp	63%	59%	7%	34%	14%	10%	17%	1%
Fscar	81%	78%	9%	48%	1%	1%	5%	0%
Sm	50%	50%	5%	40%	41%	39%	2%	37%
Espb	61%	56%	5%	44%	6%	6%	3%	4%
Ff	72%	73%	6%	61%	35%	39%	6%	16%
Cefet	88%	88%	6%	84%	45%	48%	2%	47%
Pe	87%	88%	6%	88%	41%	38%	3%	25%
Mep	87%	87%	6%	83%	67%	72%	4%	40%
Spscar	90%	90%	6%	88%	77%	72%	2%	96%
Pb	85%	86%	5%	91%	43%	46%	2%	55%
Rgs	97%	96%	6%	99%	49%	54%	3%	48%
Puc	96%	95%	6%	89%	41%	40%	1%	100%
Mg	92%	93%	6%	93%	100%	100%	3%	86%
P	100%	100%	6%	100%	36%	39%	2%	54%
Ei	81%	81%	5%	94%	35%	77%	100%	2%

Table 8.3. Fuzzy Global Productivities for 2000 and 2001 together

Course	OGOG	OGEG	OCEG	OGEC
Sc	1%	2%	100%	0%
Rj	97%	96%	13%	41%
Sp	74%	71%	7%	46%
Fscar	86%	79%	9%	43%
Sm	49%	49%	5%	34%
Espb	70%	68%	5%	60%
Ff	65%	65%	5%	52%
Cefet	86%	86%	6%	74%
Pe	85%	86%	5%	85%
Mep	85%	85%	6%	77%
Spscar	91%	91%	6%	85%
Pb	89%	89%	5%	97%
Rgs	89%	90%	5%	93%
Puc	90%	91%	6%	86%
Mg	86%	87%	5%	98%
P	100%	100%	6%	100%
Ei	80%	81%	5%	89%

of Sc in 2001 affects differently the different measures. Taking more upper boundaries as reference, *OGEC* is the more affected. Its evaluation of Ei and Sp, for instance, changes radically.

Comparing the results of 2000 and 2001 together with the results obtained treating the data of each year separately, we find a better agreement of the two years result with that of 2000. This is due mainly to the repetition in the two years data of the 2000 data of Sc, an extreme point. It is also interesting to notice that three courses, Rj, Espb and Pb, present in the analysis of the two years together better results than in each of the two years separately.

Table 8.4 presents the Malmquist indices according to DEA and to *OGOG*, which is the fuzzy concept closest to DEA. The indices are calculated on a yearly basis and from the aggregate 1998-1999 to 2000-2001. DEA results were generated using DEAP [17].

Table 8.4. Malmquist Indices

	OGOG				DEA			
Course	98/99	99/00	00/01	98+99/00+01	98/99	99/00	00/01	98+99/00+01
Sc	1.33	0.64		0.68	1.07	0.85		0.97
Rj	0.97	1.01	4.96	1.34	1.35	1.07	1.06	1.18
Sp	2.98	0.31	10.49	0.63	1.08	0.52	1.24	0.84
Fscar	1.27	1.04	0.54	1.23	1.19	1.05	1.11	1.19
Sm	0.31	2.54	0.64	1.28	1.12	1.44	1.27	1.48
Espb		0.79	3.76	1.01		0.88	1.21	0.94
Ff	2.05	0.79	0.3	0.97	1.9	0.89	0.87	1.17
Cefet		1.57	0.72	1.66		2.49	0.87	2.36
Pe		1.1	0.92	1.19		1.13	1	1.29
Mep	1.12	1.13	0.84	1.2	1	1.23	0.92	1.17
Spscar	0.55	0.94	1.12	0.77	1.74	0.92	1.05	0.64
Pb		1.04	1.81	1.19		1.11	1.15	1.1
Rgs	1.14	1.06	0.51	1.09	1.01	1	0.95	0.9
Puc	1.36	0.98	0.62	1.11	1.64	2.85	0.92	2.75
Mg	1.05	1.25	0.84	1.23	1.23	1.01	0.87	1.08
P		1.14	1.12	1.27		1.96	0.68	1.82
Ei	1.22	0.91	0.97	1.02	1.32	1.23	1.07	1.21
Mean	1.28	1.07	1.89	1.11	1.3	1.27	1.02	1.3
St. Dev.	0.68	0.46	2.62	0.25	0.3	0.61	0.16	0.54

Both approaches, but mainly the fuzzy one, applied on a yearly basis, capture the effect of variations that are corrected the next year. For instance, there is a strong effect of the reduction of Faculty size in Rj and Sp from 2000 to 2001. This reduction may reflect a permanent downsizing move but may also mean only a temporary answer to the importance given to the denominator of the productivity indices in the last triennial evaluation. It must be brought to consideration that the input column of Sp, before presenting the reduction from 25 to 17 in 2001, had presented, from 1999 to 2000, an opposite variation, from 18 to 25.

As we limit the comparison to averages of two years data, the indices become much closer to 1. The highest positive variation of relative productivity are then of Cefet, with a 1.66 fuzzy Malmquist index and a 2.36 DEA-Malmquist index, and of Puc, with a 1.11 fuzzy index and 2.75 as its DEA index. The lowest indices are of Sp with 0.63 for the fuzzy Malmquist index and 0.84 for the DEA-Malmquist index and of Spscar with 0.77 and 0.64 respectively.

All the other measures present strong oscillations in the Malmquist indices taken in a yearly basis. These oscillations occur in different units as we change the measurement concepts. Table 8.5 presents the correlation coefficients between the Malmquist indices generated according to different fuzzy points of view and of these indices with DEA-Malmquist indices. In general, the correlations increase as we take biennial data. We can notice no high correlation in Table 5, showing the ability of the different measures considered of signaling different trends according to the point of view that the analyst may wish to take.

Table 8.5. Correlations between Malmquist Indices Vectors

Measures	98/99	99/00	00/01	98+99/00+01
OCOG-DEA	.02	.39	.48	.57
ECEG-DEA	.06	.37	.17	.55
OGEG-DEA	-.18	.47	.1	.46
OCEC-DEA	.45	.33	.44	.44
OCOG-ECEG	.98	1	.22	.99
OCOG-OGEG	.84	.58	-.14	.86
OCOG-OCEC	.61	.96	-.5	.75
ECEG-OGEG	.80	.56	-.4	.86
ECEG-OCEC	.70	.97	.44	.79

8.7 Conclusion

The fuzzy approach allows comparing academic performances under different points of view. Simple model assumptions allow overcoming the absence of previous information about the probability distribution of the stochastic disturbances.

The different concepts led to different ranking. Thus, a practical result of the analysis is to show that every course in the set may reach positions of high and low efficiency depending on the concept of efficiency employed.

Other practical results can be established. Extremely high and extremely low Malmquist indices are generated when the analysis is based on the evolution of inputs and outputs from one year to the other. Much more reasonable indices appear when the evolution of two years averages is analyzed. Thus, the idea of collecting data annually but waiting two or three years to produce evaluations seems to be more sensible than producing annual evaluations.

References

1. Charnes A H, Cooper W W, Rhodes E (1978) Measuring the Efficiency of Decision Making Units. European Journal of Operations Research 2: 429-444
2. Farrell M J (1957) The measurement of productive efficiency. Journal of the Royal Statistical Society, A 120: 449-460
3. Banker R D, Charnes A H, Cooper, W W (1984) Some Models for estimating Technical and Scaling Inefficiencies in DEA. Management Science 30: 1078-1092 (1984)
4. Charnes A H, Cooper W W, Golany B, Seiford L M, Stutz J (1985) Foundations of DEA for Pareto-Koopmans efficient production functions. Journal of Econometrics 30: 91-107
5. Kumbhakar S C, Lovell C A K (2000) Stochastic Frontier Analysis. Cambridge University Press, Cambridge, U. K.
6. Simar L.,Wilson W (1998) Sensitivity analysis of efficiency scores: How to bootstrap in nonparametric frontier models. Management Science 44: 49-61
7. Morita H, Seiford L M (1999) Characteristics on Stochastic DEA Efficiency - Reliability and Probability being Efficient. Journal of the Operations Society of Japan 42: 389-404
8. Sant'Anna A P (2002) Data Envelopment Analysis of Randomized Ranks, Pesquisa Operacional 22: 203-216
9. Yamada Y, Matui, T Sugiyama M (1994) New analysis of efficiency base don DEA. Journal of the Operations Research Society of Japan 37: 158-167
10. Zadeh L (1965) Fuzzy Sets, Information and Control 8, 338-353 (1965).
11. Montgomery D C (1997) Introduction to Statistical Quality Control. J. Wiley, 3rd. ed., N. York
12. Caves D W, Christensen L R, Diewert, W E (1982) The Economic Theory of Index Numbers and the Measurement of Input, Output and Productivity. Econometrica 50: 1393-1414
13. Malmquist S (1953) Index Numbers and Indifference Surfaces. Trabajos de Estadistica 4: 209-242
14. Fare R, Grosskopf S, Lindgren B, Roos P (1989) Productivity Developments in Swedish Hospitals: a Malmquist Output Index Approach. Discussion Paper 89-3, Southern Illinois University, USA (1989)
15. Sant'Anna A P (1998) Dynamic Models for Higher Education in Various Sites. Proceedings of the ICEE-98, Rio de Janeiro-BR (1998).
16. Sant'Anna A P (2001) Qualidade, Produtividade e GED. Anais do XXXIII SBPO, C. Jordo-BR.
17. Coelli T J (1996) A Guide to DEAP Version 2.1: A Data Envelopment Analysis (Computer) Program. CEPA Working Paper 96/8, University of New England, Australia

Index

Author Index

Reviewer List

Ajith Abraham	Leandro dos Santos Coelho
Albert Y. Zomaya	Maria do Carmo Nicoletti
Andy M. Tyrrell	Maria Ganzha
Annibal Parracho Sant'Anna	Mario Borges Neto
Carlos A. Coello Coello	Mario Köppen
Christian Blum	Marley Maria B. R. Vellasco
El-Ghazali Talbi	Mary A. Cooksey
Enrique Alba	Matjaz Gams
Felipe M. G. Frana	Michael O'Neill
Jean Robillard	Miguel Redondo
Jenny Eriksson Lundström	Rosa Maria Vicari
Jordi Vallverdú	William B. Langdon

MIX
Papier aus verantwortungsvollen Quellen
Paper from responsible sources
FSC® C105338

If you have any concerns about our products,
you can contact us on
ProductSafety@springernature.com

In case Publisher is established outside the EU,
the EU authorized representative is:
Springer Nature Customer Service Center GmbH
Europaplatz 3, 69115 Heidelberg, Germany

Printed by Libri Plureos GmbH
in Hamburg, Germany